Normal and Abnormal Circadian Characteristics in Autonomic Cardiac Control: New Opportunities for Cardiac Risk Prevention

Normal and Abnormal Circadian Characteristics in Autonomic Cardiac Control: New Opportunities for Cardiac Risk Prevention

Mikhail Matveev
Rada Prokopova
and
Choudomir Nachev

Nova Science Publishers, Inc.
New York

Copyright © 2006 by Nova Science Publishers, Inc.

All rights reserved. No part of this book may be reproduced, stored in a retrieval system or transmitted in any form or by any means: electronic, electrostatic, magnetic, tape, mechanical photocopying, recording or otherwise without the written permission of the Publisher.

For permission to use material from this book please contact us:
Telephone 631-231-7269; Fax 631-231-8175
Web Site: http://www.novapublishers.com

NOTICE TO THE READER

The Publisher has taken reasonable care in the preparation of this book, but makes no expressed or implied warranty of any kind and assumes no responsibility for any errors or omissions. No liability is assumed for incidental or consequential damages in connection with or arising out of information contained in this book. The Publisher shall not be liable for any special, consequential, or exemplary damages resulting, in whole or in part, from the readers' use of, or reliance upon, this material.

Independent verification should be sought for any data, advice or recommendations contained in this book. In addition, no responsibility is assumed by the publisher for any injury and/or damage to persons or property arising from any methods, products, instructions, ideas or otherwise contained in this publication.

This publication is designed to provide accurate and authoritative information with regard to the subject matter cover herein. It is sold with the clear understanding that the Publisher is not engaged in rendering legal or any other professional services. If legal, medical or any other expert assistance is required, the services of a competent person should be sought. FROM A DECLARATION OF PARTICIPANTS JOINTLY ADOPTED BY A COMMITTEE OF THE AMERICAN BAR ASSOCIATION AND A COMMITTEE OF PUBLISHERS.

Library of Congress Cataloging-in-Publication Data
Normal and abnormal circadian characteristics in autonomic cardiac control : new opportunities for cardiac risk prevention / Mikhail Matveev, Rada Prokopova, and Choudomir Nachev, editors.
   p. ; cm.
Includes bibliographical references and index.
ISBN 1-59454-908-7
1. Heart rate monitoring. 2. Heart beat--Measurement. 3. Circadian rhythms. 4. Autonomic nervous system. 5. Cardiovascular system--Diseases--Prevention.
[DNLM: 1. Cardiovascular Diseases--physiopathology. 2. Autonomic Nervous System--physiology. 3. Autonomic Nervous System--physiopathology. 4. Circadian Rhythm--physiology. 5. Heart Rate--physiology. WG 120 N842 2006] I. Matveev, Mikhail. II. Prokopova, Rada. III. Nachev, Chudomir.
QP113.N667   2006
616.1'2--dc22
                                                                                                          2005036617

Published by Nova Science Publishers, Inc. ✤ *New York*

CONTENTS

Preface		vii
Abbreviations		ix
Chapter 1	Neurogenic Control of the Cardiac Activity *Choudomir Nachev*	1
Chapter 2	Functional Anatomy of the Cardiovascular Innervation *Choudomir Nachev*	19
Chapter 3	Heart Rate Variability and Evaluation of the Autonomic Heart Control *Mikhail Matveev*	33
Chapter 4	Circadian Characteristics of the Cardiovascular System *Rada Prokopova and Mikhail Matveev*	47
Chapter 5	Time-related Heart Autonomic Balance Characteristics in Healthy Subjects *Mikhail Matveev and Rada Prokopova*	59
Chapter 6	Changes in the Autonomic Cardiac Control in Arterial Hypertension *Rada Prokopova and Mikhail Matveev*	77
Chapter 7	Models of the Autonomic Dysbalance in Ischaemic Heart Disease *Rada Prokopova and Mikhail Matveev*	103
Chapter 8	Heart Failure and Heart Autonomic Balance: Correlating Changes in Autonomic Balance Circadian Characteristics and Ventricular Arrhythmias *Rada Prokopova and Mikhail Matveev*	127
Summary of the Results: Conclusions		145
Index		149

PREFACE

In this book we are analyzing the circadian characteristics of the autonomic cardiac control in norm and in pathology through the indices of heart rate variability. Our starting hypothesis was that the changes in the circadian nature of the autonomic balance have both specificity and sensitivity, i.e., they have a characteristic profile in the principal cardiovascular pathologies.

The material is presented in eight chapters. The first four summarize the anatomical and the functional data on the nerve control of the cardiovascular system, on the circadian nature of the different systems securing that control, and on the variability of the heart rate as a method for studying the autonomic control of the heart.

The next four chapters summarize the results of our studies clarifying the profiles of the circadian characteristics of the autonomic control in healthy individuals and in patients with arterial hypertension, ischaemic heart disease and heart failure.

The authors are grateful to

Schiller AG, Switzerland,
Schiller Engineering – Sofia, Bulgaria,
Les Laboratoires SERVIER, Bulgaria,
Roche Bulgaria Ltd,
Berlin-Chemie Sofia, Bulgaria,
Tchaikapharma Inc., Bulgaria

with whose help the research on the topic was conducted in the past five years.

The authors

ABBREVIATIONS

-A-

AB	autonomic balance
ABP	arterial blood pressure
ACE	angiotensin converting enzyme
ADP	adenosine diphosphate
AMP	adenosine monophosphate
ANS	autonomic nervous system
AP	arterial pressure
AT II	angiotensin II
AV	atrioventricular

-B-

BHAT	Beta Blocker Heart Attack
BP	blood pressure

-C-

CAD	cardiac artery diseases
CCS	Canadian Cardiovascular Society
CHF	chronic heart failure
CK-MB	cardiac muscle isoenzyme of creatine kinase
CNS	central nervous system
CRF	chronic renal failure
CVD	cardiovascular diseases
CVI	cardiovascular incidents
CVLM	caudal ventrolateral medulla

-D-

DAP	diastolic arterial pressure
dRR	differences of the R-R intervals between subsequent heart beats
dSDRR	standart deviation of dRR intervals
dMDRR	mean deviation of dRR intervals

-E-

ECG	electrocardiography; electrocardiogram
EchoCG	echocardiography
EF	ejection fraction

-F-

FFT	Fast Fourier Transformation

-H-

HAB	heart autonomic balance
HDL	high density lipoproteins
HF	heart failure
HF	high frequency
HR	heart rate
HRV	heart rate variability
HT	hendgrip test

-I-

IHD	ischaemic heart disease
IS-5-MN	isosorbide-5-mononitrate
ISAM	Intravenous Streptokinase in Acute Myocardial Infarction

-J-

JNC	Joint national committee

-L-

LD	linear discriminator
LDA	linear discriminant analysis
LF/HF	low frequency/high frequency ratio
LF	low frequency
LVH	left-ventricular hypertrophy

-M-

MDRR	mean deviation of R-R intervals
MH	mild hypertension
MI	myocardial infarction
MILIS	Multicenter Investigation of Limitation of Infarction Size
MMI	morning myocardial infarction

-N-

NMMI	non-morning myocardial infarction
NT	normotensive
NTS	nucleus tractus solitarii

-P-

PAI	plasminogen activator inhibitor
PNN50	number of R-R (N-N) intervals differing by more than 50ms
PTCA	percutaneous transluminal coronary angioplasty

-Q-

QTd	QT dispersion

-R-

R-A-A	renin-angiotensin-aldosterone
RAAS	renin-angiotensin-aldosterone system
RSA	respiratory sinus arrhythmia
RMSSD	root mean-square deviation of the differences between adjacent R-R intervals

RS	resting state
RVLM	rostral ventrolateral medulla

-S-

SA	sinoatrial
SAP	systolic arterial pressure
SCD	sudden cardiac death
SDRR	standart deviation of R-R intervals
SD	standart deviation

-T-

TIMI	Thrombolysis in myocardial ischemia
t-PA	tissue plasminogen activator
TRABI	Time-Related Autonomic Balance changes Indicator

-U-

UA	unstable angina
ULF	ultra low frequency

-V-

VLF	very low frequency
VLM	ventrolateral medulla
VM	Valsalva manoeuvre
VNS	vegetative nervous system

Chapter 1

NEUROGENIC CONTROL OF THE CARDIAC ACTIVITY

Choudomir Nachev
St. Anna University Hospital, Sofia, Bulgaria

CARDIOPULMONARY RECEPTORS: REFLEX EFFECTS

The mechano- and haemosensitive receptors are located in the cardiac atria and ventricles, in the large veins close to the heart and in the lungs. These are free, uncapsulated nerve endings of the afferent sympathetic and parasympathetic fibers innervating these structures. The receptors bear the common name of cardiopulmonary, but they differ among themselves by the character of their sensitivity to natural stimulation (mechanical and/or chemical), by their location, by the type of the nerve fibers with which they are connected (myelinated or unmyelinated), as well as of the nerves from which these fibers originate (vagus or sympathicus) [1].

The reflex responses to stimulation of the receptors occur through the efferent autonomic nerves or through humoral pathways. The effects can be inhibitory or excitatory, depending on the involved receptors.

Receptors of the Vagal Afferent Fibers

Vagal receptor endings are located in the myocardium of all cardiac cavities, in the lungs and in the sites where the large veins flow into the atria. Their distribution in the cardiac cavities is uneven. The sino-atrial (SA) and atrioventricular (AV) nodes, as well as the atria, are richer in vagal endings compared to the ventricles and the coronary vessels [2].

The cardiac vagal receptors are connected with an intense network of nerve fibers close to the heart (*heart plexus*). From them the information reaches along n. vagus to nucleus tractus solitarii (NTS) in medulla oblongata. The link with the efferent neurons of the vagus, located in nucleus dorsalis vagi, takes place in that nucleus. Moreover, nerve projections originate from NTS to nucleus ambiguus and the vasomotor centers of the brain stem, as well as to preganglionic neurons of the sympathetic nervous system in columna intermediolateralis of the spine.

Mechanosensitive Receptors

Atria

The afferent myelinated fibers of the atria have different conduction velocity (between 8 and 65 m/s). The type of the excitation is determined by the specificity of the stimulus, which in turn depends on the anatomical site of the receptor [3].

The volume of the atria is the main factor determining the level of excitability of the receptors. Therefore, they belong to the *volume-sensitive receptors* (also known as *receptors for low pressure)*.

Ventricles

Almost all vagal receptors in the ventricles are connected with unmyelinated afferent fibers. The receptors have a very low background frequency of the discharges (1-2 impulses per second) and low conduction velocity (1.2-1.5 m/s). During rest, the receptor endings are usually quiet.

The mechanoreceptors of the cardiac ventricles are located predominantly in the epicardium, whereas the haemoreceptors are distributed throughout the myocardium.

The frequency of the afferent discharges of the ventricular receptors depends on the contractility of the myocardium: increasing the contractility results in higher frequency of the discharges, while its decreasing has the opposite effect [4].

Linear correlation has been found between the level of activity of the receptors and the diastolic pressure in the left ventricle.

A paroxysmal increase of the afferent activity of the receptors is observed in the event of severe experimental blood loss, leading to strong reduction of the ventricular filling [5]. It is assumed that this paradoxical response is evoked by the combination of "empty" ventricle with increased inotropic state and that the same effect is observed in patients with vasovagal syncope.

Haemosensitive Receptors

Atria

The haemosensitive endings in the atria are with a slightly manifested activity and they are not sufficiently well identified.

Ventricles

Unlike the atria, the ventricles are abundantly innervated with haemoreceptors. The haemoreceptors are activated as a result of chemical stimuli, especially in the case of ischaemia of the myocardium [1]. The lower-posterior wall of the left ventricle is particularly rich in haemoreceptors.

Receptors of the Sympathetic Afferent Fibers

The sympathetic afferent fibers innervate the myocardium of all cardiac cavities, but are prevalent in the ventricles.

Unlike the vagal afferent nerves, the sympathetic nerves are connected with numerous receptor fields, therefore apart from the heart, an afferent sympathetic fiber receives information from the oesophagus and the bronchi as well.

The sympathetic cardiac nerves conduct the signal received to ganglion stellatum of truncus sympathicus, from where by means of rami communicantes albi and the posterior roots it reaches the spinothalamic and spinoreticular pathway in the spine. The efferent sympathetic responses are produced at the level of the brain stem, through interneuronal links and interactions with nucleus ambiguus and the vasomotor centers.

Some afferent fibers do not rise to the more rostrally located nerve centers, but through interneurons they regulate the sympathetic efferent traffic at the level of the spinal cord. Some of the afferent fibers are linked back with ganglion stellatum, thus effecting the so-called cardiocardiac reflexes.

Mechanosensitive Receptors

Atria

The afferent sympathetic mechanosensitive fibers have a low frequency of the discharges at basal conditions (1-2 impulses per second). They are activated simultaneously with the systole or diastole of the atria, as well as in the event of stretching of the atrio-ventricular valves. When the pressure in the atria rises, the sympathetic afferent endings increase the frequency of their discharges, irrespective of the cardiac phase [1].

Ventricles

The myelinated sympathetic afferent fibers in the ventricles are more numerous than those in the atria. Their activation corresponds to the changes in the intraventricular pressure. They respond to the mechanical stimuli corresponding to the cardiac cycle, therefore their discharges have a rhythm that is similar to the cardiac rhythm.

The unmyelinated afferent sympathetic fibers are also sensitive to the changes in the ventricular pressure, but they do not demonstrate a lasting correlation in time with the cardiac cycle.

Haemosensitive Receptors

Some of the afferent sympathetic fibers in the cardiac muscle (predominantly unmyelinated) possess a selective sensitivity to chemical stimuli. The rest (both unmyelinated and myelinated) are polymodal, i.e., they respond both to mechanical and to chemical stimuli. Chemical stimulation increases the sensitivity to mechanical stimuli.

Most unmyelinated fibers are quiet or demonstrate minimal spontaneous activity.

Ventricles

The sympathetic haemoreceptors, unlike the vagal ones, are evenly distributed in the anterior and in the lower-posterior part of the left ventricle. The excitatory sympathetic response to coronary obstruction is the same, irrespective of the localization of the ischaemic lesion [6].

Reflex Responses Mediated through the Vagal Receptors

Responses Originating from the Atria

It is assumed that the vagal afferent fibers originating from the atria, the ventricles and the lungs have a tonic inhibitory effect on the vasomotor center of medulla oblongata, which normally maintains the tone of the peripheral vessels. In experimental animals the dilatation of the atria induces a vasodilatory response in the peripheral vessels with lowering of the vascular resistance.

The atrial vagal receptors are at the basis of the cardiopulmonary reflex [1]. The reflex consists in reflex tachycardia indused by rapid infusion of liquids, and it is due to distension of the atria. Cardiopulmonary reflex may also be induced by means of balloon dilation at the sites of the venous-atrial junctions. The graded cooling of the vagal nerve, which is used for interrupting the conduction of the signal along the myelinated fibers, without blocking the unmyelinated fibers (with low frequency of the discharges), shows that the afferent signals provoking the tachycardiac response, are conducted along unmyelinated vagal afferent fibers [7].

The efferent pathway of the tachycardiac response induced by distension of the atria follows predominantly the sympathetic nerves, but the final effect depends on the base levels of the sympathetic tone and on the activity exercised by the vagus on the sinus node. The influence of the reflex is limited on the sinus node and does not affect the ventricular contractility.

The cardiopulmonary reflex is not influenced by the baro- and haemoreceptor reflexes, which suggests that it possesses independent pathways originating from the atrial receptors [1].

Atrial receptors also play a role in the regulation of the volume of the extracellular fluid. The reflex responses to dilatation of the atria have a diuretic and natriuretic effect.

The plasma levels of vasopressin are also influenced by the dilatation of the atria. Activation of the atrial receptors is found to reduce the level of the circulating vasopressin, provided the other factors influencing vasopressin secretion are controlled. Such factors can be the changes in the haemodynamics and in the activity of the autonomic nervous system (ANS). The release of vasopressin is conducive to the intensification of the reflex diuresis induced by the dilatation of the atria.

Reflex Responses Originating from the Ventricles

The afferent vagal signals originating from the ventricles exercise a sympathetic inhibitory effect. Their interruption leads to considerable rise in the systemic arterial pressure, as well as to increase in the direct efferent sympathetic traffic to the muscles and to the splanchnic area, as a result of which the vascular resistance in the cited areas increases. The deactivation of the vagal receptors in the ventricles (e.g., in the case of severe blood loss or when low pressure is applied in the lower half of the body, leading to lowering of the diastolic pressure in the left ventricle), causes excitation of the sympathicus.

Activation of the vagal receptors upon rise of the pressure in the cardiac ventricles (increase of the blood volume, narrowing or obturation of the aorta) induces a vasodepressor response. Experimental studies have shown that stimulation of the vagal afferent fibers in the left ventricle leads to a general reduction of the vascular resistance, which results above all from the reflex dilatation of splanchnic vessels. Direct electrical stimulation of type C unmyelinated nerve fibers induces a strong vasodilatory response in the renal and muscle vessels [5]. Other studies using less selective stimuli confirm these data. They indicate that the activation of the cardiopulmonary receptors (including the ventricular receptors) induces reduction of the systemic vascular resistance and that this response is mediated through interruption of the efferent sympathetic activity [1]. The interruption or the deactivation of the cardiopulmonary receptors with vagal afferentation has the opposite effect and leads to a rise in the systemic and regional vascular resistance.

The change in the activity of the cardiopulmonary receptors (to which the ventricular vagal nerve endings also belong) does not have a direct chronotropic effect. These receptors are assumed to have an indirect influence on the heart rate by modulating the action of the arterial baroreceptors. These interactions probably play a role in some pathological states [1].

Being a part of the cardiopulmonary receptors, the ventricular receptors contribute in the highest degree to the effecting of the vasomotor control. This fact is of particular importance in humans on account of their characteristic erect posture and the frequent changes in the posture, leading to fluctuations in the central blood volume and cardiac filling. The activity of the cardiopulmonary receptors in these conditions changes permanently and this emphasizes their role for the haemodynamic adaptation. The correct activity of these receptors is of particular importance for the orthostatic regulation of blood pressure.

The effect of the cardiopulmonary receptors on the tone of the venous vessels and the venous capacity is less manifested compared to the effect on the arterial vascular resistance. The effect on the veins in the splanchnic area is higher.

Damage to the cardiopulmonary mechanoreceptors is at the basis of some frequently occurring morbid states. They are an important part of the mechanisms of the sympathetic inhibitory response during a vasovagal syncope.

Cardiopulmonary receptors play a role in the renin secretion as well. Their activation leads to reduction of the renin secretion. The speed with which this reflex occurs suggests that it is nerve-mediated. The severing of the afferent vagal fibers by vagotomy or freezing induces increased renin secretion. Increased renin secretion is also observed with deactivation of the cardiopulmonary receptors caused by reduced central volume, e.g., in the event of severe blood loss.

The role of the cardiopulmonary receptors for the release of vasopressin is insignificant.

Reflex Responses, Mediated through the Sympathetic Receptors

The sympathetic cardiac receptors provoke pressor responses due above all to increased sympathetic activity and to a lesser extent – to inhibition of the vagal tone. The increased sympathetic activity influences both the heart and the peripheral circulation.

Ischaemia of the myocardium is accompanied by changes in the activity of the sympathetic receptors. The mechanoreceptors may be influenced by the dilatation of the ventricles or by the changes in the regional contractility. In the event of myocardial ischaemia, the haemoreceptors are activated due to the increased release of bradykinin and anaerobic metabolites. Adenosine also participates in this process as a natural stimulus for the sympathetic haemoreceptors.

CARDIAC REGULATION AT THE LEVEL OF THE PERIPHERAL AUTONOMIC SYSTEM

The peripheral part of the autonomic nervous system (ANS) is the "bridge" between the ANS of the heart and its higher cerebral regulation at different levels in the central nervous system (CNS). The balance between the sympathetic and the parasympathetic activity is maintained by the coordinated activity of certain structures in the CNS, which secure in this way the normal functioning of the heart and the cardiovascular system.

Parasympathetic Innervation of the Heart

The afferent projections of the parasympathicus, originating from the heart, the cardiopulmonary baro- and haemoreceptors and the arterial receptors reach the CNS along cerebro-cranial nerves IX and X.

The cell bodies of the parasympathetic cardiac afferent nerves are located in ganglion nodosum, which is in foramen jugulare at the base of the skull.

The afferent information is transmitted in medulla oblongata to the efferent preganglionic parasympathetic neurons innervating the heart. Their cell bodies are located in nucleus dorsalis vagi, nucleus ambiguus and the area linking them. The neuronal projections from these nuclei move back to the heart along n. vagus. The parasympathetic innervation of the heart involves three ramifications of n. vagus on either side. On the right side, the cardiac ramifications of n. vagus join the deep heart plexus.

The axons of the efferent preganglionic neurons end in the cells of the parasympathetic ganglia in immediate proximity to the heart, on the surface of the epicardium or in the very cardiac muscle.

Sympathetic Innervation of the Heart

The information from the sympathetic receptors of the heart and the aorta reaches the CNS along the spinal nerves. The cell bodies of the sympathetic afferent cardiac nerves are in the ganglia of the dorsal roots of the spinal nerves, between the segments C6 and Th6 of the spine.

The efferent preganglionic neurons of the cardiac sympathicus are located in columna intermediolateralis of the upper thoracic part of the spine. Unlike the parasympathetic ganglia, the cardiac sympathetic ganglia form a ganglionic chain truncus sympathicus, located far from the heart. The axons of the efferent preganglionic sympathetic neurons of columna intermediolateralis in the upper thoracic segments of the spine move along the Th_1-Th_5 roots to the cranial pole of g-n stellatum, the median cervical ganglion and the mediastinal ganglia.

The sympathetic efferent fibers leaving the cited ganglia in the direction of the heart form an upper, middle and lower cervical and thoracic sympathetic cardiac nerve.

Along the entire efferent sympathetic system of the heart there exist numerous feedback mechanisms including excitatory and inhibitory processes. The modulation of the secretion of noradrenaline from the postganglionic sympathetic endings is controlled by such mechanisms [8].

The atrial sympathetic nerve fibers innervate predominantly the tissue of the atria, but they also penetrate in the ventricles, while the influence of the ventricular sympathetic nerves is restricted to the actual ventricles.

Cardiac Autonomic Plexus

After entering the chest, the efferent vagal and sympathetic fibers join in the heart plexus, which represents an intricate network of mixed nerves and contains both preganglionic vagal and postganglionic sympathetic fibers, and afferent nerves from the cardiopulmonary receptors.

The heart plexus consists of a surface and a deep part. The left upper cervical sympathetic nerve joins the surface heart plexus, while the remaining sympathetic cardiac nerves are directed to the deep heart plexus.

The sympathetic nerves are located along the surface of the myocardium, while the parasympathetic nerves penetrate deep into the subendocardial layers at the base of the ventricles, from where they disperse to the remaining subendocardial areas. In the mesocardiac layers of the ventricular tissue there is overlapping of the two types of efferent autonomic nerve fibers: sympathetic and parasympathetic [9].

The ramifications of the cardiac plexus along the coronary arteries are known as right and left coronary plexus [8,9].

Anatomical and functional symmetries exist in the autonomic innervation of the heart. The parasympathetic and the sympathetic cardiac nerves have a parallel course, but those on the right are directed predominantly to the sinus node (exercising antagonistic influences on its chronotropic function), while the nerves on the left are directed mainly to the atrioventricular node and the ventricles [10]. The innervation from the left affects the atrioventricular conduction, the threshold for the appearance of ventricular fibrillation, the duration of the QT-interval, as well as the ST-segment of the ECG [11]. These anatomical and physiological specificities are of an important clinical significance.

Both the sympathetic and the parasympathetic efferent preganglionic neurons in the CNS are under the constant control of higher cerebral representations of the cardiac regulation. The harmonious interaction of these structures secures optimum effectiveness of the cardiac muscle during each cardiac cycle.

The autonomic control of the vascular tone occurs mainly through sympathetic efferent influences, while the heart rate and the cardiac stroke volume depend on the balance between the sympathetic and the parasympathetic impacts [12].

CARDIORESPIRATORY CENTERS OF MEDULLA OBLONGATA

The centers for cardiovascular control function in close interaction with those of respiration. The stem mechanisms securing the homeostasis of the cardiovascular and respiratory functions constitute a common system for cardiac and respiratory control [13,14]. Medulla oblongata is the place where the afferent information and the motor efferent neurons of that system meet. NTS is the information relay of the system [13].

The information originating from the baroreceptors of the heart and the blood vessels is transported along cerebro-cranial nerves IX and X to NTS in its caudal part. The first synapses of the baroreceptor reflex occur in NTS. Lesions or functional the disordered in the activity of the NTS neurons influence both the level of the arterial pressure (AP), and the rhythm of the cardiac activity [14,15,16].

In addition to NTS, cell representations of the system for cardiac regulation are also found in the serotoninergic nuclei of raphe, the parabrachial area, the periaqueduct gray matter, the hypothalamus, the amygdale and the cerebral cortex [9,17].

The ventrolateral medulla (VLM) has a key role in the control of the cardiac activity and of the vascular tone. Being a component of the baroreceptor reflex arc, it receives information from the carotid baroreceptors through the vagal afferent fibers and NTS. The information that is processed in the VLM and is then directed to other stem structures is of critical importance to AP regulation.

Two separate areas of VLM participate in the regulation of arterial pressure (AP): rostral (RVLM) and caudal (CVLM).

The RVLM neurons are tonically active and the level of their activity determines the basal sympathetic tone.

RVLM is the center of the tonic reflex regulation of AP. Chemical stimulation of its neurons provokes extreme rises of AP. The increased RVLM activity is at the basis of the neurogenic forms of arterial hypertension.

Most of the RVLM cells are adrenergic (cell group C_1), but some use other neurotransmitters, such as glutamate, encephalin, substance P, neuropeptide Y [14,18]. Microinjection of L-glutamate in RVLM causes a considerable rise in the blood pressure in hypertensive rats and a weaker but still significant rise in the pressure of normotensive animals. The heart rate is not significantly affected in any of the groups [19].

In the case of arterial hypertension the RVLM neurons receive an abnormally big number of excitatory impulses. This leads to increased sympathetic activity, which underlies the development and maintenance of the mechanisms of this disease [15].

Some antihypertensive drugs influence directly the RVLM neurons, and more specifically their imidazole, serotonin and alpha-2-adrenergic receptors.

RVLM regulates the activity of the preganglionic sympathetic neurons in the thoracic-lumbar part of the spine, which affects the release of catecholamines from the medullar part of the suprarenal gland [20], as well as on the cardiac production of atrial natriuretic factor [21].

The disordered in the sympathetic control of the cardiac activity in the case of hyperfunction of RVLM can be an important pathogenic factor and can contribute to the development of cardiac hypertrophy and ischaemia, congestive heart failure, as well as to the emergence of arrhythmias and myocardial infarction. This emphasizes the great clinical significance of the level of the sympathetic cardiac tone, the pathways for its regulation starting at the level of the brain stem [14].

NTS exercises inhibitory influences on RVLM, their result being reduction of the sympathetic tone and of AP.

Episodes of uncontrolled hypertension and tachycardia are observed in patients with disordered function of the baroreflex due to deafferentation of the baroreceptors. In the case of bilateral NTS lesions and preserved RVLM there is severe damage of the baroreflex, manifested with exacerbated pressor responses and sharp drops of AP after food or after intake of alpha-2-agonists with central action [22].

CVLM has inhibitory functions. Stimulation of its neurons leads to vasodepressor responses. The tonic inhibitory control on the basal sympathetic tone starts from the CVLM.

CVLM has a fundamental significance for the effecting of the baroreceptor reflex. It conducts the baroreceptor impulses received rostrally to RVLM, controlling in this way its activity.

The elimination or the weakening of the activity of the CVLM neurons may cause arterial hypertension.

Catecholamine Groups in the Ventrolateral Medulla

In VLM there are cell groups composed of catecholamine neurons: noradrenergic group A_1 and adrenergic group C_1.

The noradrenergic group A_1 is located in close proximity to CVLM and it possesses a vasodepressor action, which is partly is connected with inhibition of the vasopressin secretion.

The adrenergic group C_1 coincides to a certain extent with the CVLM structures and like them it has vasodepressor action.

The cell groups A_1 and C_1 form a column along the entire ventrolateral medulla oblongata. Group A_1 forms the lower part of that column, which sends projections to NTS, the parabrachial complex and the hypothalamus. Group C_1 represents the upper part of the column and participates in the regulation of the activity of the preganglionic sympathetic neurons through efferents to NTS, locus coeruleus and the hypothalamus.

The interactions of RVLM and CVLM, on the one hand, and the neurons from group A_1 and C_1 – on the other are not completely clear. The nucleus of n. facialis is also a part of the vasomotor structures of the ventrolateral medulla.

Organization of the Baroreceptor Reflex at Medullar Level

The neuronal local circles in medulla oblongata are the basis for the occurrence of the baroreceptor reflex. Other levels of the central nervous system (the insular cortex, the central amygdaline nucleus and the hypothalamus) also have an important participation in the total integration of the reflex, but the neuronal population forming the center of the circle responsible for the baroreceptor reflex function is located in medulla oblongata.

This circle is formed of excitatory and inhibitory neurons reciprocally interconnected with synaptic links in which glutamate and gamma-aminobutyric acid are used as neurotransmitters. The circle reacts to transient changes in the blood pressure by the negative feedback mechanism, with the corresponding effects on the heart rate and on the peripheral resistance.

Briefly, the baroreceptor impulses originating from the aortic arc and from the carotid bifurcation, through ramifications of n. vagus and n. glossopharingeus, stimulate the baroreceptor sensory neurons of NTS. The NTS cells (second neuron) transmit the baroreceptor impulses in the depressor area of CVLM. The CVLM neurons send projections to RVLM, where they have a direct inhibitory action. In this way, the circle is suppressed.

The excitability of the RVLM neurons is regulated tonically by inhibitory neurons. By lowering the increased excitability that RVLM induces in the preganglionic sympathetic neurons, the baroreceptor stimulation causes activation of the inhibitory neurons of CVLM, which leads to bradycardia and vasodepressor effect.

Inhibitory Preganglionic Vagal Neurons in Medulla Oblongata

The inhibitory efferent preganglionic vagal neurons of medulla oblongata are located in the ventral medullar reticular formation (80%) and the dorsal vagal nucleus (20%) [23].

SUPRABULBAR REGULATION OF THE HEART

Amygdale

The amygdale and the hypothalamus are important subcortical stations forming a bridge between the cortical and the lower lying stem and spinal structures. The afferent and the efferent links between these structures outline probable nerve pathway along which the regulation of the cardiac rhythm takes place [17].

Studies in recent years demonstrate the important role of the amygdale for the cardiovascular reflexes and especially for those accompanying emotions [17].

The amygdale is composed of numerous subnuclei, the central one of which has a key role in the processing of the autonomic responses. A large number of projections originating from the insular and orbitofrontal cortex, from the parabrachial nucleus and from NTS are directed to it. Experimental studies in animals and in humans during surgical treatment of some forms of epilepsy demonstrate that the central amygdaline nucleus receives afferent baroreceptor impulses and that the activity of its neurons correlated with the phases of the cardiac cycle [24,25].

The changes in the heart rate, induced by stimulation of the amygdale, are not accompanied by changes in the blood pressure and do not depend on them. Stimulation of the central and the lateral nucleus of the amygdale, of the parvocelular part of the basal nucleus, of the paraamygdaline complex and of the putamen cause bradycardia and delayed respiration. The effects of the stimulation of the magnocellular part of the basal nucleus are opposite. The tachycardia is sympathetically mediated, while the bradycardia is due to parasympathetic activation and inhibition of the sympathicus [23].

Stimulation of the median nucleus of the amygdale provokes cardiac arrhythmias. At first, there is prolongation of the QT-interval, changes in the ST-segment and increase of the amplitude of the T-wave. These ECG changes persist even after discontinuation of the stimulation and lead after a certain time to bradycardia, sometimes followed by idioventricular rhythm and ventricular fibrillation.

The amygdale is the principal center for integrative control of the cardiac activity and of the autonomic functions. The rich afferent and efferent links of this nucleus show that it is subjected to numerous different impacts. A large number of nerve fibers originate from the central nucleus of the amygdale (and also from the nucleus of the stria terminalis), which are directed to the stem autonomic centers. They are part of the neuronal circles participating in the regulation of the cardiac rhythm, in which influences from the cerebral cortex are also concentrated [23,26].

Hypothalamus

The impulses generated in the cardiovascular receptors are directed to the hypothalamus, from which numerous powerful efferent influences start back to the brain stem.

The electrical stimulation of the anterior hypothalamus provokes bradycardia and arterial hypotension, while the irritation of the lateral and posterior hypothalamus leads to the onset of tachycardia and rise in the blood pressure. The nature of the responses from the anterior and from the posterior hypothalamus is also determined by the basal level of the vagal and sympathetic tone [23].

Numerous links exist between the lateral hypothalamus and the insular lobe of the cerebral cortex, through which the chronotropic impacts on the heart upon microstimulation of the insular cortex are effected. It has already been proven with certainty that precisely the lateral hypothalamus is the place where the synapses of the pathway of the tachycardiac responses to insular stimulation take place, and that glutamate is the neurotransmitter mediating the synaptic transmission [27].

Stimulation of the hypothalamus frequently results in cardiac arrhythmias. Apart from the sinus, ventricular tachycardia is also observed, preceded by prolongation of the QT-interval and a remarkable increase of the T-wave. The gradually increasing intensity of the stimulations of the anterior thalamus results first in generation of ventricular extrasystoles and ventricular tachycardia and finally – ventricular fibrillation. Bilateral vagotomy has no effect on these responses, but they are influenced by propranolol, which indicates that they are due to activation of the sympathetic nervous system [16].

The cardiovascular effects of the hypothalamus occur through hypothalamic projections directed to the periaqueduct gray matter of the mesencephalon, to the stem reticular substance and to columnae intermediolaterales of the spine.

Lateral hypothalamus

This part of the brain is known above all for its role in the regulation of the motivation, feelings and the concomitant autonomic symptoms. This is an inhomogeneous area, which resembles the reticular formation in structure. It is connected above all with the limbic cortex. The insular fields with hypertensive effect are better represented through their projections in the lateral hypothalamus than those with hypotensive effect [23].

The lateral hypothalamus is connected also with the gray matter of the periaqueduct, with the parabrachial nucleus and with nuclei of the reticular substance, with the dorsal vagal complex and with the spine.

Paraventricular Hypothalamus

The paraventricular nucleus of the hypothalamus has descendent projecting fibers, which link it directly with the preganglionic sympathetic neurons in columnae intermediolaterales of the spine. Projections of the central nucleus of the amygdale are also added to them in descendent direction, linking the paraventricular hypothalamus with some stem nuclei that are important for the cardiovascular regulation: the gray matter of the periaqueduct space, the parabrachial nucleus, parts of the reticular substance of medulla oblongata, nucleus ambiguous and nucleus dorsalis n. vagi. In this way, the paraventricular hypothalamus may influence the preganglionic neurons both of the sympathicus, and of the parasympathicus. The paraventricular hypothalamus receives rich afferentations from the limbic cortex.

The descendent projections from the amygdale and the hypothalamus form three main pathways originating accordingly from: (i) the amygdale and the nucleus of the stria terminalis, (ii) the lateral hypothalamus, and (iii) the paraventricular hypothalamus. They conduct the impacts originating from numerous different cortical areas, which are finally concentrated on the structure uniting all cortical and subcortical representations of the cardiovascular regulation: the periaqueduct gray matter [23]. On its part, this structure is in close interaction with the autonomic nuclei of the brain stem.

Cortical Areas Participating in the Cardiac Regulation

The first cardiac responses to stimulation of the brain were observed in 1875 (after [23]). They show that the stimulation of the motor cortex is accompanied by tachycardia, without concomitant changes in the blood pressure. Later changes in the cardiac activity were also registered upon stimulations of the sigmoid cerebral cortex, the frontal part, the subcallous gyrus, the septal field, the temporal lobe and gyrus cinguli.

Changes in the electrocardiogram (ECG) are most frequently observed. Deepening of the Q-tooth, changes in the magnitude and polarity of the T-wave and of the QRS-complex, and elevation or depression of the ST-segment are described. These phenomena are more frequent upon stimulation of the sigmoid gyrus. During stimulation of the hippocampus, they are usually accompanied by prolongation of the QT-interval. In some cases the changes in the ECG appear many hours after discontinuation of the stimulation, sometimes also with ventricular fibrillation.

In recent years, the improvement of the research methods made it possible to broaden substantially the knowledge on the cardiac control exercised by the structures located high in the central nervous system.

Insular Cortex

This part of the cerebral cortex is located on the claustrum. Areas were found to exist in the posterior part of the insula, whose stimulation induces tachycardia and arterial hypertension, while stimulation of neuronal areas located in the vicinity leads to hypotension and bradycardia. It has been found that the fields in the left insula inducing tachycardia are located in the more superficial layers of the posterior cortex, while the bradycardiac fields located deeper, with a certain overlapping of the two types of fields [28]. This chronotropic map of the left insular cortex is the first demonstration of a cortical-cerebral field with specific cardiac representation. A series of impulses, 80 to 100 milliseconds before the peak of the T-wave – the time during which the ventricular myocardium is least stable – induces a progressively deepening AV-block, leading to total cardiac block, interventricular block, prolongation of the QT-interval, depression of the ST-segment, ventricular ectopy and finally – asystolic death [29]. The ECG changes were accompanied by cardiac myocytolysis. Haemorrhages are observed in the areas adjacent to the origin of the left branch of the His bundle. Parallel to this, a rise in the plasma levels of noradrenaline was observed and hence it is assumed that the cardiac changes stem from a rise in the sympathetic activity.

A rise in the blood pressure with tachycardia was registered in patients subjected to partial resection of the temporal lobe due to treatment-resistant epilepsy as a result of the stimulation of the anterior insular lobe, whereby the responses are more persistent in the case of stimulation to the right than to the left. Hypotensive responses combined with bradycardia are provoked more easily in the cases of stimulation of the left insula [30].

These experimental results contributed to the identifying in the cerebral cortex (pallium) of fields with chronotropic impact on the cardiac activity, which may generate cardiac arrhythmias upon prolonged stimulation.

All earlier data indicate that the insular cortex has a determining role in the regulation of the cardiac activity. This is in support of the theory, according to which the insula participates in the modulation of the frequency and the rhythm of the heart in states of emotional stress, especially bearing in mind also its numerous links with the limbic system [17]. The insula is assumed to participate in the pathogenesis of the arrhythmias in patients with cardiac damage [23].

Other Cortical Areas

Cardiovascular responses are also registered after stimulation of the motor cortex. The interruption of the pyramidal pathway in the area of medulla oblongata provoked their disappearance. In addition to intact pyramidal pathway, these responses also require preservation of the integrity of the nerve pathway connecting the motor cortex with the hypothalamus.

Changes in the cardiac activity are also observed upon stimulation of the prefrontal cortex and septum. There is evidence of chronotropic cardiac representation in gyrus cinguli, in the lower part of the temporal lobe and in uncus hippocampi.

The cardiac responses induced by stimulation of the orbito-frontal cortex are characterized with bradycardia and slight lowering of the blood pressure. Injection of small quantities of the patient's own blood into the same cerebral area induces marked ECG changes, manifested in large T-waves, depression of the ST-segment, ventricular ectopies, and sometimes also malignant cardiac arrhythmias.

As the orbitofrontal area and the adjacent areas of the limbic cortex are connected with numerous structures regulating the autonomic nervous system, they are assumed to participate also in the pathogenesis of these responses. Some nuclei of the thalamus and the lateral hypothalamus, as well as the parabrachial nucleus, are such autonomic structures. In addition to processing the incoming autonomic information and conducting it to the orbitofrontal cortex, they are connected in descending direction with the preganglionic sympathetic neurons located in columnae intermediolaterales of the thoracolumbar part of the spine.

According to some authors, sympathetic excitation only is not sufficient for provoking malignant arrhythmias and sudden death upon stimulation of the cited cortical areas. A simultaneous rise in the parasympathetic tone is also necessary [31].

The sites in the brain, which seem to play an especially important role in the regulation of the balance of the autonomic nervous system, are the frontal lobe, the insular cortex and the amygdale [17,31]. Simultaneous blocking of the amygdale and frontal cortex is in a position to normalize the elevations of the arterial pressure in all experimental models of hypertension. It is assumed that such a block could protect the individual against dangerous descendent central-nerve influences, leading to bilaterally increased autonomic tone [31].

Periaqueduct Gray Matter

All cortical and subcortical fields described so far send their projections to the gray matter of the periaqueduct space.

The efferent projections of the gray matter of the periaqueduct space are directed mainly to RVLM – an area controlling the sympathetic excitatory impulses to the cardiac preganglionic sympathetic neurons.

The numerous projections passing through the gray matter of the periaqueduct, which are responsible for the connection between the cerebral cortex and RVLM, participate in the pathogenesis of arrhythmias under stress [23].

***

The cortical, amygdaline and hypothalamic structures participating in the control of the cardiac activity are identified. Hierarchic organization exists between them, and each can influence the activity of the axis located lower in the in nerve (Figure 1.1).

The cardiac responses effected at the level of the brain stem are the fastest upon stimulation of the baroreceptors. The amygdale is involved in a higher integration of these reflexes, including and the emotions, thus the cardiovascular responses correspond to the emotional content of different situations in life. The insular cortex, and probably other cerebral cortical fields as well, such as the orbitofrontal cortex and gyris cinguli, modulate the responses integrated from an emotional point of view, by potentiating or weakening them depending on the experience of the individual and on the concrete circumstances.

The insular and orbitofrontal cortex are the principal cortical centers for regulation of the heart. They are a part of the limbic system and they function in close interaction with it, which emphasizes the exceptionally important physiological role of this system in the control of the cardiac activity, especially in emotive states and stress [17].

The susceptibility of the heart to the arrhythmogenic effects of the nerve axis insula-amygdale-hypothalamus probably changes under pathological conditions. There exists a hypothesis according to which in the course of the development of the cardiac diseases, the changes occurring in the cardiac receptors result in changes in the afferentation with remodelling of the synaptic links in the central nervous system and abnormal intensification of the efferent responses of the autonomic nervous system – e.g., of those which originate from the insula.

The disordered in the balance between the activity of the neurons of the insula, the amygdale and the hypothalamus lead to changes in the cardiac activity and arrhythmias. Sudden death is an "electrical incident" caused by fatal cardiac arrhythmias. Cerebrogenic sudden death, as well as non-fatal cardiac arrhythmias, may occur after strong emotional excitation or may complicate the course of all types of stroke, epileptic seizures, cerebrocranial traumas and neurosurgical procedures [32].

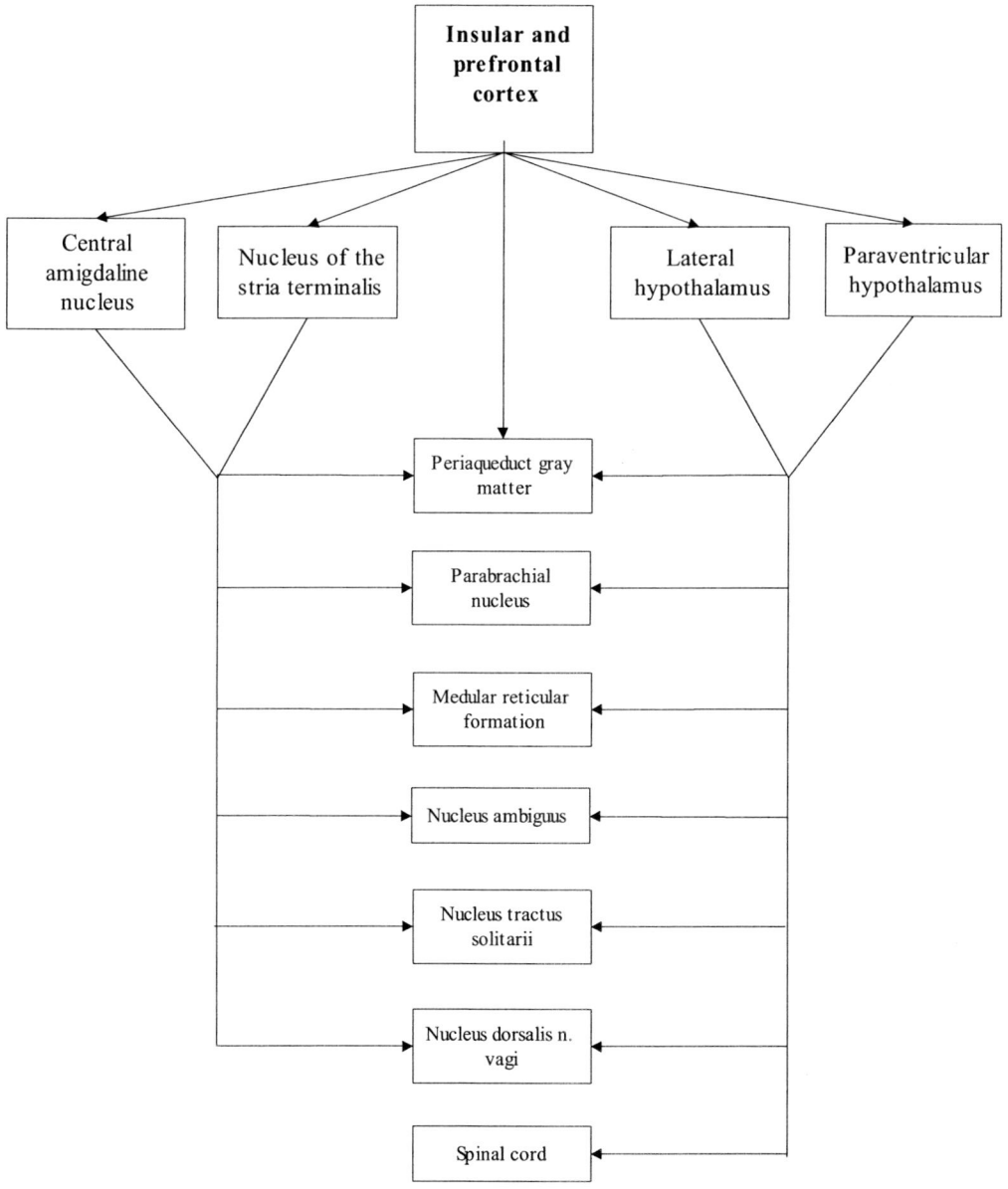

Figure 1.1. Basic cortex and subcortex centres for cardiac activity control.

REFERENCES

[1] Smith, ML; Thames, MD. Recettori cardiaci: caratteristiche di scarica ed effetti riflessi. In: Armour, JA; Ardell, JL, editors. *Neurocardiologia*. Roma: CIC Edizioni Internazionali; 1996; 13-40.

[2] Loffelholz, K; Pappano, AJ. The parasympathetic neuroeffector junction of the heart. *Pharmacol Rev*, 1985, 37, 1-24.

[3] Hainsworth, R. Reflexes from the heart. *Physiol Rev*, 1991, 71, 617-658.

[4] Gupta, BN; Thames, MD. Behavior of left ventricular mechanoreceptors with myelinated and nonmyelinated afferent vagal fibers in cats. *Circ Res*, 1983, 52, 291-301.

[5] Oberg, B; Thoren, P. Circulatory responses to stimulation of medullated and non-medullated afferents in the cardiac nerve in the cat. *Acta Physiol Scand*, 1973, 87, 121-132.

[6] Mimisi, AJ; Thames, MD. Distribution of left ventricular sympathetic afferents demonstrated by reflex responses to transmural myocardial ischemia and to intracoronary and epicardial bradykinin. *Circulation*, 1993, 87, 240-246.

[7] Linden, RJ; Mary, DASG; Weatherill, D. The effect of cooling on transmission of impulses in vagal nerve fibres attached to atrial receptors in the dog. *Q J Exp Physiol*, 1981, 66, 321-332.

[8] Armour, JA. La regolazione cardiaca a livello del sistema nervoso autonomo periferico. In: Armour, JA; Ardell, JL, editors. *Neurocardiologia*. Roma: CIC Edizioni Internazionali; 1996; 187-208.

[9] Levine, SR, Patel, VM; Welch, KMA, Skinner, JE Are heart attacks really brain attacks? In: Furlan, AJ, editor. *The heart and stroke*. London-Berlin-Heidelberg-New York-Paris- Tokyo: Springer Verlag; 1987; 185-216.

[10] Levy, MN; Warner, MR. Effetti del parasimpatico sulla funzione cardiaca.In: Armour, JA; Ardell, JL, editors. *Neurocardiologia*. Roma: CIC Edizioni Internazionali; 1996; 41-61.

[11] Naver, HK; Blomstrand, C; Wallin, G. Reduced heart rate variability after right-sided stroke. *Stroke*, 1996, 27, 247-251.

[12] Barnes, KL; Ferrario, CM. Role of the central nervous system in cardiovascular regulation. In: Furlan, AJ, editor. *The heart and stroke*. London-Berlin-Heidelberg-New York-Paris-Tokyo: Springer Verlag; 1987; 155-169.

[13] Cheng, Z; Quo, SAJ; Lipton, DC. Domoic acid lesions in nucleus of the solitary tract: time-dependent recovery of hypoxic ventilator/ response and peripheral afferent axonal plasticity. *J Neurosci*, 2002, 22, 3215-3226.

[14] Hopkins, DA; Ellenberger, HH. I centri cardiorespiratori del midollo allungato: relazione tra efferenze ed afferenze. In: Armour, JA; Ardell, JL, editors. *Neurocardiologia*. Roma: CIC Edizioni Internazionali; 1996; 237-262.

[15] Colombari, E; Sato, MA; Cravo, SL; Bergamaschi, CT; Campos, CT Jr; Lopes, OU. Role of the medulla oblongata in hypertension. *Hypertension*, 2002, 38, 549-556.

[16] Phillips, AM, Jardine, DL; Parkin, PJ; Hughes, T; Ikram, H. Brain stem stroke causing baroreflex failure and paroxysmal hypertension. *Stroke*, 2000, 31, 1997-2001.

[17] Ter Horst, GJ. Central autonomic control of the heart, angina, and pathogenic mechanisms of post-myocardial infarction depression. *Eur J Morphol*, 1999, 27, 257-266.
[18] McLean, KJ; Jarrot, B; Lawrence, AJ. Hypotension activates neuropeptide Y-containing neurons in the rat medulla oblongata. *Neurosci*, 1999, 92,1377-1387.
[19] Bergamaschi, C; Campos, RR; Schor, N; Lopes, OU. Role of the rostral ventrolateral medulla in maintenance of blood pressure in rats with Goldblatt hypertension. *Hypertension*. 1995, 26, 1117-1120.
[20] Ross, CA; Ruggiero, DA; Park, DH; Joh, TH; Sved, AF; Fernando-Pardal, J; Saavedra, JM; Reis, DJ. Tonic vasomotor control by the rostral ventrolateral medulla: effect of electrical and chemical stimulation of the area containing C1 adrenaline neurons on arterial pressure, heart rate, plasma catecholamines and vasopressin. *J Neurosci*, 1984, 4, 474-494.
[21] Jiao, JH; Guyenet, PG; Baertschi, AJ; Lower brain stem controls cardiac ANF secretion. *Am J Physiol*, 1992, 263, H198-H207.
[22] Biaggioni, I; Whetsell, WO; Nadeau, JH. Baroreflex failure in a patient with central nervous system lesions involving the nucleus tractus solitarii. *Hypertension*. 1994, 23, 491-495.
[23] Oppenheimer, SM; Hopkins, D. Regolazione soprabulbare del cuore. In: Armour, JA; Ardell, JL, editors. *Neurocardiologia*. Roma: CIC Edizioni Internazionali; 1996; 263-291.
[24] Frysinger, RC; Harper, RM. Cardiac and respiratory correlation with unit discharge in epileptic human temporal lobe. *Epiepsia*, 1990, 31, 162-171.
[25] Zhang, J-X; Harper, PM; Frysinger, RC. Respiratory modulation of neuronal discharge in the central nucleus of the amygdala during sleep and waking states. *Exp Neurol*, 1986, 91, 193-207.
[26] Holstege, G; Meiners, L; Tan, K. Projections of the red nucleus of the stria terminalis to the mesencephalon, pons, and medulla oblongata in the cat. *Exp Brain Res*, 1985, 58, 379-391.
[27] Butcher, KS; Cechetto, DF. Receptors in lateral hypothalamic area involved in insular cortex sympathetic responses. *Am J Physiol*, 1998, 275, H689-H696.
[28] Oppenheimer, SM; Cechetto, DF. Cardiac chronotropic organization of the rat insular cortex. *Brain Res*, 1990, 533, 66-72.
[29] Oppenheimer, SM; Wilson, JX; Guiraudon, C; Cechetto, DF. Insular cortex stimulation produces lethal cardiac arrhythmias: a mechanism of sudden death? *Brain Res*, 1991, 550, 115-121.
[30] Oppenheimer, SM; Gelb, AW; Girvin, JP; Hachinski, VC. Cardiovascular effects of human insular stimulation. *Neurology*, 1992, 42, 1727-1732.
[31] Skinner, JE. Brain/Heart. Discussion. *Stroke*,1993, 24, Suppl. 1, 10-12.
[32] Cheung, RT; Hachinski, V. The insula and cerebrogenic sudden death. *Arch Neurol*, 2000, 57, 1685-1688.

Chapter 2

FUNCTIONAL ANATOMY OF THE CARDIOVASCULAR INNERVATION

Choudomir Nachev
St. Anna University Hospital, Sofia, Bulgaria

CARDIAC INNERVATION

Cardiac innervation is oriented towards certain structures of the heart and is characterized by being lateral to the innervating pathways. The right sympathetic and vagal nerves are directed more to the sinus node, while the left innervate predominantly the AV-node.

The stimulation of the right g-n stellatum induces sinus tachycardia and has a much weaker effect on the conduction in the AV-node, while the stimulation of the left g-n stellatum usually causes shifting of the sinus guide of the rhythm to an ectopic place and it shortens substantially the time for the conduction and the refractoriness in the AV-node, but does not increase always the frequency of the discharges in the sinus node.

The stimulation of the right cervical vagal nerve leads predominantly to delay of the frequency of the discharges in the sinus node, while the effect of the stimulation of the left vagus is manifested mainly in prolongation of the conduction time and the refractoriness in the AV-node.

Due to the selective innervation, the autonomic nerves to the sinus node can be severed without affecting the innervation of the AV-node, and conversely – the vagal or the sympathetic nerves of the AV-node can be severed without triggering changes in the sinus innervation [1].

Vagal control predominates in some cardiac structures (an example of this is the sinus node and its automatism), while in other structures, notably the ventricles, the influences of the sympathicus dominate [2]. The autonomic nerve effects may be manifested in a different way even within one cardiac structure. For example, the atria and especially the SA-node are permanently subjected to antagonistic impacts on the part of the vagal and sympathetic nerves. This antagonism is obvious after blocking of one or the other nerves, whereby only the specific impacts of the unblocked nerves are residual. Blocking of the vagal nerves increases the heart rate. Inhibition of the sympathetic activity lowers the frequency. The permanent impacts of the vagal and sympathetic nerves determine their *tone*. Since the rhythm of the totally denervated heart (the autonomic cardiac rhythm) is considerably faster than the heart rate during rest, it is assumed that during rest the tone of the vagal nerves predominates over tone of the sympathetic nerves.

Both sympathetic and vagal stimulation do not affect the normal conductivity in the His bundle, but both types of stimulation influence the abnormal conductivity in the AV-node [1,2].

Hypersensitivity to the respective transmitters develops after vagal or sympathetic denervation in the denervated structures.

The control of the cardiac activity is relevant to the frequency of the cardiac contractility (chronotropic action), the force of the contractility (inotropic action) and to the velocity of the atrioventricular conduction (dromotropic action).

Chronotropy. The stimulation of the right vagal nerve or the direct impact on the SA-node with acetylcholine leads to lowering of the heart rate (negative chronotropic effect). With strong impacts even stopping of the heart is possible. The stimulation of the sympathetic nerves or the impact with noradrenalin leads to accelerated cardiac rhythm (positive chronotropic effect). The action of the vagal nerves prevails during simultaneous stimulation of the sympathetic and the vagal nerves. The sympathetic nerves increase the automatism of all parts of the conducting system of the heart, hence upon inhibition of the leading pacemaker it depends precisely on these nerves how quickly the function of pacemaker will be assumed by a secondary pacemaker and how effective its action will be. What is more, the sympathetic nerves have a positive chronotropic impact on the cells of the pacemaker whose spontaneous activity was suppressed by exogenous factors. Under the influence of these nerves, the activity of the ectopic excitation foci may grow and hence the risk of appearance of arrhythmias may increase.

Inotropy. The changes in the cardiac rhythm in itself influence considerably the force of the contractility. Moreover, the cardiac nerves influence directly the contractility. Upon impact from the vagal nerves, the force of the contractility of the atria decreases (negative inotropic effect). Upon impact from the sympathetic nerves, the intensity of the contractility both of the atria and of the ventricles increases (positive inotropic effect).

Dromotropy. In norm the cardiac nerves influence the conduction of excitation only in the area of the AV-node. The sympathetic nerves stimulate the atrioventricular conduction and thus shorten the interval between the contractions of the atria and the ventricles (positive dromotropic effect). Under the effect of the vagal nerves (especially the left), the atrioventricular inhibition increases, even to complete transient AV-blocking (negative dromotropic effect).

Vagal Innervation

The vagal ganglia of the heart are in immediate proximity to it, which is characteristic of the vagal innervation of all remaining organs as well. Some of the cardiac vagal ganglia are located in special fatty deposits around the ostium of the large veins.

As was noted already, the right and the left vagal fibers influence in different ways the functions of the sinus and AV-node, whereby the stimulation of the right nerves causes considerable sinus bradycardia, and with increased activity of the left vagal fiber a tendency for the development of an AV-block is observed.

In experiments with dogs it has been demonstrated that vagal stimulation in the area of the fatty deposits around the right pulmonary veins controls the bilateral vagal influences on the sinus node, while the interruption of the vagal fibers does not affect the control on the AV-node. Conversely, the vagal ganglia in the fatty deposits around the ostium of the lower vena cava and the lower part of the left atrium are important predominantly for the vagal influences on the AV-node and the removal of that tissue has no effect on the vagal stimulation of the sinus node.

Sympathetic Innervation

Similar to the vagal nerves, the sympathetic nerves on the right have a greater influence on the sinus node and the heart rate, while the left sympathetic nerves exercise a greater influence on the conductivity of the AV-node.

Specificities of the Autonomic Innervation of the Cardiac Structures

Sinus node

The sinus node is richly innervated with adrenergic (sympathetic) and cholinergic (parasympathetic, vagal) postganglionic nerve endings, but it is above all under parasympathetic control [2]. The highest concentration of vagal nerves in the heart is in the sinus node, where accordingly the highest concentrations of acetylcholine, acetylcholine esterase and cholineacetyl transferase. By releasing acetylcholine, vagal stimulation delays the frequency of the discharges in the sinus node and can sometimes cause blocking at the output of the node. Acetylcholine increases, while noradrenalin reduces the refractoriness in the center of the sinus node. Adrenergic stimulation increases the frequency of the sinus discharges. The vagal action on the sinus node is manifested with initial negative dromotropic effect, followed by a negative chronotropic effect [2]. The phase of the cardiac cycle in which the vagal discharge appears and the basal sympathetic tone have a substantial significance for the vagal impacts on sinus frequency and conductivity.

Atria

Vagal influences dominate in the atria. After the sinus node, the concentration of acetylcholine and of the enzymes connected with it is the highest in the atria (more in the right than in the left). Vagal stimulation reduces the duration of the action potential of the atrium. Due to the inhomogeneity of the vagal innervation of the atria, a strong vagal stimulation may cause dispersion of the refractoriness of the atria, resulting in atrial fibrillation.

Atrioventricular node

The AV-node (as well as the area of the His-bundle) are richly innervated by cholinergic and adrenergic fibers whose density exceeds that in the ventricular myocardium. Ganglia, nerve fibers and nerve networks (plexuses) are located in immediate proximity to the AV-node.

The AV-node is distinguished with rich vagal innervation which is inferior only to that of the sinus node. The stimulation of the vagus, especially of the left vagal nerve, increases the time for conduction in the AV-node, as well as the refractory period.

With sympathetic stimulation – both right and left – the time for conduction in the AV-node decreases, and the refractory periods become shorter. The activity of additional rhythm pacemakers can be potentiated in this way.

Ventricles

The ventricles have an abundance of sympathetic nerve fibers. The left ventricle is innervated predominantly by the left sympathetic nerves, although there is a certain overlapping. Intraindividual differences also exist. It is assumed that the misbalance between the right and the left sympathetic innervation of the ventricles participates in the pathogenesis of arrhythmias in the syndrome of the long QT-interval [3]. It can be said most generally that the right sympathetic chain shortens the refractoriness of the anterior wall of the ventricles, while the left influences predominantly the posterior wall, although there are fields of overlapping.

The existence of considerable vagal innervation of the ventricles as well, which probably also participates in the arrhythmogenesis, was proven in the past 2-3 decades. The acetylcholine concentration in the ventricles is relatively low and does not exceed 20 to 50% of that in the atria [1].

The ventricular vagal fibers are located deep in the wall of the ventricle, passing along it into the subendocardial layer. Some of the vagal fibers directed to the septum use other pathways, probably along the His-Purkinje system [2].

The sympathetic innervation of the ventricles follows pathways that differ from those of the vagus. Functional and anatomical studies show that the afferent and the efferent sympathetic nerves follow the surface of the epicardium before cutting in depth to innervate the endocardium. In the surface layers of the epicardium the sympathetic nerve fibers follow the course of the coronary arteries.

The differences in the location of the sympathetic and vagal innervation of the heart allow to understand better the mechanisms of the arrhythmias occurring after cardiac ischaemia and infarction. For example, vagal and not sympathetic denervation can be assumed in subendocardial infarction. At the same time, transmural infarction, by severing the sympathetic nerves, may lead to sympathetic denervation of myocardial areas far from the place of the ischaemia [4]. The sympathetic deafferentation plays an important role for the occurrence of cardiac arrhythmias [2].

His-Purkinje system and ventricular myocardium

The stimulation of the adrenergic beta-receptors reduces the refractoriness of the His-Purkinje system and of the myocardium of the ventricles, increasing the slope of the diastolic depolarization in phase 4 [2]. Vagal stimulation has a weaker effect that the sympathetic one on the electrophysiological properties of the myocardium by increasing the refractory period in the His-Purkinje system and in the ventricular myocardium. It is assumed that the vagal effect is influenced by the modulating effect of the stimulation of the beta-receptors, while the stimulation of the alpha-2-receptors suppresses the presynaptic release of noradrenalin.

Effects of the Autonomic Cardiac Innervation

Effects of the vagal stimulation

The vagus has a modulating effect on the cardiac sympathetic activity at pre- and postsynaptic level by regulating the quantity of the released noradrenalin and by inhibiting induced by the cyclic adenosine monophosphate (AMP) phosphorylation of the cardiac proteins.

The neuropeptides released from the nerve fibers of the two branches of the ANS also have a modulating effect on the autonomic responses. For example, the neuropeptide Y released from the sympathetic terminals suppresses the cardiac vagal effects [1,2].

Tonic vagal stimulation causes greater reduction of the sinus frequency, if it is applied against the background of tonic basal sympathetic stimulation. This sympathetic-parasympathetic interaction is known as *accentuated antagonism*.

Unlike the effects on the sinus node (and the sinus frequency), the changes in the AV-conductivity with simultaneous sympathetic and vagal stimulation are the sum of the individual responses of AV-conductivity to separately applied vagal and sympathetic stimulation [1].

The cardiac responses to brief vagal stimuli start after a short latency period and attenuate quickly. Unlike them, the cardiac responses to sympathetic stimulation start and attenuate slowly.

The rapid onset with rapid attenuation of the responses to vagal stimulation allows dynamic beat-to-beat vagal modulation of the heart rate and AV-conductivity. Conversely, the delayed response to sympathetic stimulation does not allow it to affect the beat-to-beat regulation of the sympathetic activity [1].

Periodic vagal activation (which occurs every time when the systolic wave of the pressure reaches the baroreceptor areas in the aorta and the carotid sinus) induces changes in the duration of the sinus cycle, as a result of which the sinus node releases faster or slower discharges for the periods of time identical to those of the vagal discharges. The vagal discharges prolong in a similar way the conduction time in the AV-node, moreover being influenced by the basal levels of the sympathetic tone. Since the peak vagal effects on the sinus frequency and the conductivity of AV-node are manifested at different times of the cardiac cycle, a brief vagal discharge may delay the sinus frequency, without affecting the AV-conductivity, or conversely – it could prolong the conduction time in the AV-node without delaying the sinus frequency [1].

The vagal nerves exercise minimal but measurable effects on the ventricular tissue by reducing the intensity of the myocardial contraction and by prolonging the refractoriness. Under certain circumstances, acetylcholine can have a positive inotropic effect. Vagal stimulation has indirect impacts as well, modulating the sympathetic effects. The afferent vagal activity appears to be higher in the posterior ventricular myocardium. This may be of significance for the vagomimetic effects of the lower myocardial infarction [1].

Effects of the sympathetic stimulation

The stimulation of the sympathetic ganglia shortens the refractory period in the epicardium and in the underlying endocardium of the left ventricular wall. Dispersion of the regenerative properties is observed, i.e., a difference is found in the degree of shortening of the refractoriness in the different areas of the epicardium. The uneven distribution of the electrophysiological effects is explained to a certain extent with the uneven distribution of noradrenalin, because the ventricular content of noradrenalin is higher in the base than in the apex of the heart. Besides, the muscle cells are richer in noradrenalin than the Purkinje cells [1,2].

Interactions between the Sympathicus and the Vagus

From everything said so far it becomes clear that it is not possible to describe the effects of one part of the ANS without taking into effect the influences of the other. The interaction between the sympathicus and the vagus is manifested both at pre- and at postsynaptic levels in CNS, and at the level of the peripheral nerves.

As a rule, the two ANS components have antagonistic actions with respect to the cardiac activity: stimulation of one part of the ANS usually suppresses the activity of the other. An example in this respect is the regulation of the sinus automatism and the atrioventricular conductivity. Nevertheless, the activities of the sympathicus and of the vagus are synergistic in some tissues, e.g., the effect on the refractory period of the atria, where both the vagal and the sympathetic impacts are reduced to lowering of the refractoriness [2].

A third variant of the interaction is observed when due to weak basal activity of one of the two ANS components there is no marked antagonism on the part of the other [2].

There exists a fourth variant as well, whereby the interaction between the two ANS components is manifested in the same was as with the accentuated antagonism. An example of accentuated antagonism is the more pronounced reduction of the heart rate upon stimulation of the vagus, against the background of stronger sympathetic activity (i.e., the vagal effect is more pronounced at a high basal sympathetic activity) [5]. Besides, although the vagus has a very weak influence on the duration of the ventricular action potential, the absolute refractory period and the threshold of the ventricular fibrillation, vagal activation with simultaneous stimulation of the sympathicus is already in a position to limit and modulate the sympathetic effect, which is also a manifestation of accentuated antagonism [2].

Takahashi and Zipes [6] have documented the phenomenon of accentuated antagonism by demonstrating that the stimulation of the vagus increases the duration of the cycle of the sinus node and the conduction time of the AV-node, but the effect is more pronounced if and the sympathicus is stimulated simultaneously. It is also shown that direct stimulation of the sympathicus contributes much more to that phenomenon than the infusion of noradrenalin or isoprenaline.

A manifestation of the antagonism between the two ANS components is the fact that the activation of the vagus suppresses the release of noradrenalin, and the stimulation of the sympathicus inhibits the release of acetylcholine.

Denervation Hypersensitivity

The denervation hypersensitivity is a phenomenon characterized by inadequately strong response of the denervated tissues to the action of certain agents. Hypersensitivity of the denervation occurs in the heart when the circulating catecholamines react with a live but sympathetically denervated myocardial tissue.

The denervation hypersensitivity creates a predilection for the appearance of cardiac arrhythmias. A correlation is found between the sympathetic denervation, the denervation hypersensitivity and the increased arrhythmogenesis after myocardial infarction [2]. These data indicate that the sympathetic denervation is very frequently arrhythmogenic and that the combination of ischaemic myocardial lesion with regional sympathetic denervation represents a substrate with a high arrhythmogenic potential. This has been confirmed in other studies as well, which show that regional sympathetic denervation in the case of transmural infarction facilitates the onset of ventricular arrhythmias [7].

The serious consequences that the sympathetic denervation of the heart could have evoke a justified interest in the mechanisms of its development in the conditions of myocardial ischaemia.

REGULATION OF THE VASCULAR SYSTEM

Vascular Tone

Many vessels have a definite quantity of smooth-muscle cells, which shrink periodically spontaneously. These contractions do not depend on the nerve influences and exist even in the case of denervation of the vessels. Owing to this phenomenon, the walls of the vessels are in a state of tension even during rest, i.e., they have a *myogenic basal tone*. The tension of most of the vessels during rest is caused both by the basal tone and by the contractions of the smooth-muscle cells, induced by the impulses coming from the vasoconstrictive vegetative fibers. The summary tension of the vascular wall is known as vascular *tone during rest*.

Vegetative Regulation of the Regional Blood Circulation

The adaptation of the blood in the peripheral vessels is controlled both by local, and by humoral and nerve mechanisms. The influence of these mechanisms on the smooth muscles of the vessels in different organs is different. The mechanisms can have a synergistic or antagonistic action on the vascular tone.

The nerve regulation of the vascular lumen is performed by the vasoconstrictive nerves of the sympathetic part of the vegetative system. Irrespective of this, parasympathetic fibers also participate in some vascular responses. The vegetative nerves innervate all blood vessels with exception of the capillaries. The density and the functional significance of this innervation vary considerably in the different organs and in the different parts of the vascular system. Noradrenalin is a mediator of the vasoconstrictive nerves, which always induces contraction of the vascular muscles.

Influence of Adrenaline and Noradrenalin

The two hormones have a generalized action on the vascular muscles. While noradrenalin is the principal transmitter of the vasoconstrictive nerves, adrenaline is the main hormonal agent. The vascular reaction to the action of the two substances is different. Adrenaline has both vasoconstrictive and vasodilating effect. Besides, the muscle reaction of the different vessels to adrenaline is different depending on their sensitivity to it. The divergent influences of the catecholamines in the blood on the vascular muscles are due to the two adrenoreceptors: alfa- and beta-receptors. The excitation of the alfa-receptors leads to contraction of the muscles; the excitation of the beta-receptors – to their relaxation. Noradrenalin acts predominantly on the alfa-receptors, and adrenaline – on the alfa- and beta-receptors. Both types of receptors exist in the blood vessels, although their quantitative ratio is different in the different parts of the vascular system. If the alfa-receptors predominate in the vessels, then adrenaline induces their shrinking, if the beta-receptors predominate – they are dilated. It is also necessary to bear in mind that the excitatory threshold of the beta-receptors is lower than that of the alfa-receptors, although the effects of the alfa-receptors predominate with the excitation of both. In this way, at low (physiological) concentrations, adrenaline induces dilation of the vessels, and at high concentrations– their narrowing.

The action of the catecholamines in pathological states is different. In normal myocardial vessels the stress and the discharge of adrenergic transmitters leads to vasodilation, in atherosclerotic vessels – to vasoconstriction. Such abnormal reaction is also observed in the collateral-dependent myocardium. The probable reason for this reaction is the prevalence of the alpha-receptors in the small vessels of the heart. Beta-receptors prevail in the healthy and in the mature collateral vessels, leading to vasodilation [8].

Renin-Angiotensin System

Excreted in the blood, the enzyme renin fragments angiotensinogen to angiotensin I. Under the action of the plasma converting enzyme, angiotensin I is transformed into angiotensin II. Angiotensin II has a very strong direct vasoconstrictive effect on the arteries and to a lesser extent – on the veins. Besides, it excites the central and the peripheral sympathetic structures, resulting in a rise of the peripheral resistance and blood pressure. Angiotensin II is also the chief stimulator for the production of aldosterone. Aldosterone increases the sodium concentration in the organism and the extracellular fluid. Simultaneously it raises the excitability of the smooth vascular muscles upon impact of vasoconstrictive agents, intensifying in this way the pressor action of angiotensin II. Due to these close interrelations between renin, angiotensin and aldosterone, they are frequently united in a renin-angiotensin-aldosterone system (RAAS), which plays an important role for the normalization of the blood circulation in the event of pathological lowering of the arterial pressure and/or the blood volume.

AUTONOMIC BALANCE OF THE HEART AND CARDIOVASCULAR DISEASES

The autonomic balance plays an important role in the regulation of the cardiovascular system when cardiovascular diseases (CVD) develop. The deviation of the balance in the direction of sympathetic hyperactivity may cause different CVD or may appear when they develop.

Cardiac arrhythmias may also cause damage to the ANS by influencing directly the electrophysiological properties of the cardiac tissue, notably the generation, conduction and refractoriness of impulses.

In the case of essential hypertension the hyperactivity of the sympathetic part was established a long time ago. In borderline, mild or early hypertension, many authors find increased sympathetic tone that leads to higher cardiac frequency, beat volume and spasm of the vessels [9]. Other authors indicate abnormal response in the progeny of hypertensive individuals to sympathetic stimulation compared to the response in the progeny of normotensive individuals [10]. According to Petretta, Bianchi, Marchiano, et al. [11], the patients with already developed left-ventricular hypertrophy (i.e., hypertension with affected target organs) have a strongly deteriorated autonomic balance. Increased sympathetic tone also leads to activation of the renin-angiotensin system, which has a direct influence on the development of the left-ventricular hypertrophy, hypertrophy of the vascular wall and the higher degree of hypertension [11]. *These finds presupposed that the disordered vegetative balance is responsible for the appearance and development of organ complications in hypertensive disease.*

In ischaemic heart disease (IHD) the disordered autonomic balance plays a role both in the acute and in the chronic stage. Many studies have identified the link between the hypersympathicotony and the poorer prognosis after myocardial infarction (MI) [12]. Other authors have studied the uncompicated IHD and have proven a damaged parasympathetic tone in it [13]. A connection is also seen between hypersympathicotony and the development of atherosclerosis [14].

ANS also has indirect effects on the heart through changes in the general haemodynamics, local blood flow and metabolism. The myocardial ischaemia is sometimes arrhythmogenic in itself, but it can also cause arrhythmias due to disorders in the autonomic control of the cardiac function.

The sympathetic and the vagal innervation of the heart follow different pathways, hence the effects of the focal myocardial ischaemia and myocardial infarction on the vagal and sympathetic afferent and efferent nerves are also different.

In heart failure (HF) the sympathicus is strongly activated as a compensatory mechanism against the low arterial pressure and the disordered perfusion of vitally important organs, caused by the worsed left-ventricular function. Subsequently, this hypersympathicotony leads to even more deteriorated perfusion and additional left-ventricular dysfunction. The patients with the highest catecholamines levels have the poorest prognosis [15]. These data suggest that *the degree of disturbance of the autonomic balance has a prognostic role.*

It is also necessary to take into account the influence of the sympathetic nervous system on the kidney, because the regulation of the arterial pressure is closely connected with the activation of the renin-angiotensin system and partially with the reabsorption of sodium and water in the case of sympathetic hyperactivity. This triggering of yet another humoral system plays an important role in hypertension, in the case of aggravate of the hypertensive disease and when left-ventricular hypertrophy develops. The increased activity of the renin-angiotensin system accelerates the thickening of the muscle layer of the intima of the arteries and increases the oxidative stress in atherosclerosis; in ischaemia there is a link with the remodelling of the ventricles; in heart failure – with the sodium and water inhibition.

THE STIMULATION TESTS

The tests involving suppression or stimulation of the sympathicus or the vagus with drugs or with physiological methods are known as stimulation. The specific impact on one of the two parts of the autonomic nervous system through the stimulation test should be understood conditionally: this is *predominantly impact on one part in* a *degree allowing to ignore the involvement of the other.*

Parasympathetic Stimulation: Valsalva Manœuvre (VM)

This test traces the frequency response of the heart during and after the test. The person examined blows into a modified mercury manometer, maintaining a pressure of 40 mm Hg for of 15 seconds. The changes in the pulse frequency are registered during the experiment and after it through ECG recording.

The test was reported for the first time by Antonio Maria Valsalva in 1704. In 1851 Weber reports about changes in the pulse induced with the test. In 1920, Flack modified the manoeuvre by proposing blowing against a mercury column and by describing the complex physiological changes accompanying the stimulation. The changes are identified in four phases by from Hamilton and his co-workers (see, e.g., in [16, 17]). The *first phase* is manifested immediately at the beginning of the blowing, causing a sudden rise in the intrathoracic pressure, with elevation of the arterial pressure as well. With continuing effort – *second phase* – the venous reflux is limited, the cardiac capacity drops, the arterial blood pressure is lowered, with subsequent rise in the pulse rate. *The third phase* occurs with interruption of the respiratory pressure and during the next several heart beats there is an ongoing decline of the cardiac capacity and drop in the blood pressure in connection with an increase of the venous pulmonary capacity after easing of the increased intrathoracic pressure. *The fourth phase* is the period before the return of the circulation to a normal state. A new rise in the blood pressure is observed, caused by the increased cardiac capacity, since the systemic vascular resistance continues to be increased in response to the lowered arterial pressure in the second phase. This repeated rise in the arterial pressure causes reflex bradycardia.

The reflex changes in the Valsalva manoeuvre are complex and they involve both the vagal and the sympathetic conduction pathways. Pharmacological studies have shown that the heart rate response can be eliminated with atropine blocking, hence this test provokes predominantly parasympathetic response. The changes in the arterial pressure have a more complicated genesis, therefore when the Valsalva manoeuvre is applied, only the changes in the pulse frequency are registered. The test is most frequently used for studying cardiac vegetative parasympathetic damage.

Sympathetic Stimulation: Handgrip Test (HT)

This test uses a manual dynamometer for exercising pressure (by folding the palm), equal to 30% of the possible maximum effort for the person tested. The effort lasts for 4-5 minutes, the arterial pressure being registered every minute. The pulse is also registered with ECG recording. The static effort maintained is a form of stress in response of which a rise in the arterial pressure is observed, and to a lesser extent – a rise in the heart rate as well. The mechanism of the rise in the arterial pressure is believed to be of reflex nature: it is assumed to be a reflex originating from the tense muscles. The ischaemic effect in the tense muscles leads to metabolic changes in them, and the latter activate the efferent pathways from the involved muscles. The elevation of the pressure is partly determined by the increase in the minute volume, due to the accelerated pulse rate, by the increase in the myocardial contractility and by peripheral vasoconstriction (which is not detected by all authors) [17]. The effort of the muscle groups is connected with increase of the sympathetic tone and rise in the level of the plasma catecholamines [18].

The static muscle effort is considered to be an easily applicable and easily reproducible test for studying the sympathetic part of the ANS [19].

From a practical point of view it is important to note that Valsalva manoeuvre and the handgrip-test are convenient physiological procedures provoking different parts of the autonomic nervous system [20]. The effect of the stimulation of the vagal and sympathetic parts is traced well through the heart rate variability (HRV) indicators, therefore in our studies we apply both tests for evaluation of the autonomic balance.

REFERENCES

[1] Zipes, DP. Genesis of cardiac arrhythmias: Electrophysiological considerations. In: Braunwald, E, editor. *Heart disease: A textbook of cardiovascular medicine. Part II: Normal and abnormal circulatoy function*. Chapter 20. Philadelphia: W.B. Saunders Company; 1998; 548-592.

[2] Mitrani, R; Zipes, D. *Neurocardiologia* clinica: le aritmie. In: Armour JA, Ardell L editors. *Neurocardiologia*. Roma: CIC Edizioni Internazionali;1996; 313-339.

[3] Schwartz, PJ; Locati, E; Moss, AJ; Crampton, RS; Trazzi, R; Ruberti, U. Left cardiac sympathetic denervation in the therapy of the congenital long QT-syndrome: a worldwide report. *Circulation*, 1991, 84, 503-511.

[4] Inoue, H; Zipes, DP. Time course of denervation of efferent sympathetic and vagal nerves after occlusion of the coronary artery in the canine heart. *Circ Res*, 1988, 62, 1111-1120.

[5] Levy, MN. Sympathetic-parasympathetic interactions in the heart. *Circ Res*, 1971, 29, 437-445.

[6] Takahashi, N; Zipes, DP. Vagal modulation of adrenergic effects on canine sinus and atrioventricular nodes. *Am J Physiol*, 1983, 244, H775-H781.

[7] Herre, JM; Wetstein, L; Lin,YL; Mills, AS; Dae, M; Thames, MD. Effect of transmural versus nontransmural myocardial iinfarction on inducibility of ventricular arrhythmias during sympathetic stimulation in dogs. *J Am Coll Cardiol*, 1988, 11, 414-421.

[8] Witzleb, E. Functions of the vascular system. In: Schmidt RF and Thews G. *Human Physiology*. Berlin, Heidelberg, New York: Springer-Verlag; 1983; 101-190.

[9] Minatoguchi, S; Ito, H; Ishimura, K; Suzuki, T; Tonai, N; Mori, M; Hira, S; Fujiwara, H. Plasma adrenaline modulates alfa-1-adrenoceptor mediated pressor responses and the baroreflex control in patients with borderline hypertension. *Blood Press*, 1995, 4, 105-112.

[10] Corti, R; Binggeli, C; Sudano, I; Spieker, LE; Wenzel, RR; Luscher, TF; Noll, G. The beauty and the Best: Aspects of the Autonomic Nervous System. *News Physiol. Sci.*, 2000, 15, 125-129.

[11] Petretta, M; Bianchi, V; Marchiano, F; Themistoclakis, S; Canonico, V; Sarno, D; Lovino, G; Bonaduce, D. Influence of left ventricular hypertrophy on heart period variability in patients with essential hypertension. *Journal of Hypertension*, 1995, 13, 1299-1306.

[12] Farrel, TG; Bashir, Y; Cripps, T; Malik, M; Poloniecki, J; Bennett, ED; Ward, DE; Camm, AJ. Riskc stratification for arrhythmic events in postinfarction patients based on heart rate variability, ambulatory electrocardiographic variables and signal-averaged electrocardiogram. *J Am Coll Cardiol*, 1991, 18, 687-607.

[13] Wennerblom, B; Lurje, L; Tygesen, H; Vahisalo, R; Hjalmarson, A. Patients with uncomplicated coronary artery disease have reduced heart rate variability mainly affecting vagal tone. *Heart*, 2000, 83, 290-294.

[14] Huikuri, HV; Jokinen, V; Syvanne, M; Nieminen, MS; Airaksinen, KEJ; Ikaheimo, MJ; Koistinen, JM; Kauma, H; Kesaniemi, AY; Majahalme, S; Niemela, KO; Frick, H. Heart Rate Variability and Progression of Coronary Atherosclerosis. *Arteriosclerosis, Thrombosis and Vascular Biology*, 1999, 19, 1979-1985.

[15] Thomas, JA; Marks, BH. Plasma norepinephrine in congestive heart failure. *Am J Cardiol*, 1978, 41, 233-243.

[16] Piha, SJ. Autonomic responses to the Valsalva manoeuvre in healthy subjects. *Clin Physiol*, 1995, 15, 339-347.

[17] Mateika, JH; Demeersman, RE; Kim, J. Effects of lung volume and chemoreceptor activity on blood pressure and R-R interval during the Valsalva manoeuvre. *Clin Auton Res*, 2002, 21, 24-34.

[18] Nami, R; Bianchini, C; Aversa, AM; Gragnani, S; Papini, L; Peruzzi, C; Lucani, B; Perone, AF; Johnson, S; Gennari, C. Circulating levels of catecholamines and cyclic AMP during the handgrip test in borderline hypertension. *J Cardiovasc Pharmacol*, 1986, 8, Suppl 5, S142-S144.

[19] Ewing, DJ; Irving, JB; Kerr, F; Wildsmith, JAW; Clarke, BF. Cardiovascular responses to sustained handgrip in normal subjects and in patients with diabetes mellitus: a test of autonomic function. *Clin Sci Mol Med*, 1974, 46, 295.

[20] Assessment: Clinical Autonomic Testing. Report of the Therapeutics and Technology Assessment Subcommittee of the American Academy of Neurology. *Neurology*, 1996, 46, 873-880.

Chapter 3

HEART RATE VARIABILITY AND EVALUATION OF THE AUTONOMIC HEART CONTROL

Mikhail Matveev
Centre of Biomedical Engineering, Bulgarian Academy of Sicences, Bulgaria

BASICS OF HEART RATE VARIABILITY MEASUREMENT AND ANALYSIS

Enormous material of facts has been accumulated, suggesting a clear link between the state of the vegetative nervous system and the cardiovascular pathology. These facts are an incentive for studies connected with determination of quantitative indices for evaluating the state of the vegetative nervous system (VNS). The heart rate variability (HRV) appears to be the most promising method in this respect.

At the same time, the popularity of the method raises a number of problems as well. The HRV indices are used rather frequently in support of unfounded conclusions, false interpretations and generalizations that go too far.

The justified interest in HRV triggered the appearance of numerous studies based on different methodological platforms for analysis of the variability. The *"Heart Rate Variability. Standards of Measurement, Physiological Interpretation, and Clinical Use. Task Force of the European Society of Cardiology and the North American Society of Pacing and Electrophysiology"* [1] occupies a special place among them. This work ought to be assessed as an exceptional contribution to the standardization of the quantitative methods for HRV analysis. Due to the frequent reference to this document, it will be cited in the text below as *Task Force....*

The HRV constitutes the changes in the duration of the interval between two successive heart beats. The successive intervals are measured between the R-teeth in adjacent QRS-complexes of a continuously recorded electrocardiogram – the so-called R-R intervals (Figure 3.1). Indeed, the beginning of every cardiac cycle on the electrocardiogram, connected with the excitation of the sinus node is the beginning of the P-wave. However, the R-teeth are most accurately detected on the signal on account of their high amplitude in the standard leads from the limbs. Therefore, the analysis of HRV is based on changes in the R-R intervals, and not in the P-P intervals.

Different terms are used in the literature when the changes in the duration of the cardiac cycles are studied, namely: variability of the duration of the cycle, variability of the R-R intervals (or RR-variability), etc. Recently the term "heart rate variability" became most extensively used. It is important that all terms used cites as the object of analysis precisely *the duration of the R-R intervals, and not the heart rate*. As the evaluation of HRV is performed only on successive intervals between normal QRS-complexes, then the term N-N intervals (from norm-norm) is used to designate the R-R intervals. *Normal QRS-complexes are understood to mean the complexes initiated with impulses from the SA-node.*

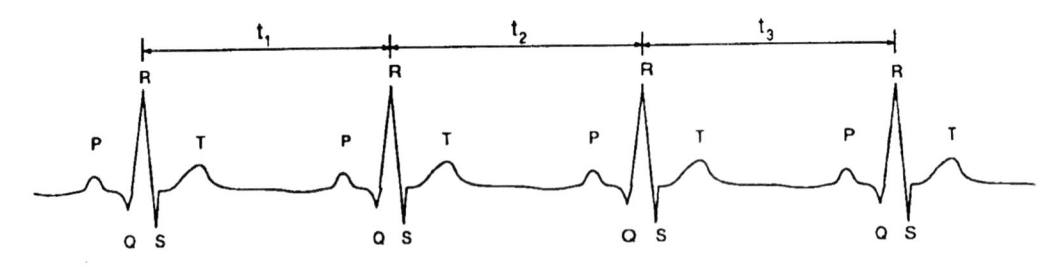

Figure 3.1. R-R intervals.

As the sequences of R-R intervals represent a time series from a mathematical and statistical point of view, their analysis is performed using time-domain and frequency-domain indices.

Two sets of indices are used in the *time-domain* to describe HRV. The first set consists of statistical evaluations of the sequence of R-R intervals (in ms) – Figure 1. This sequence is known as RR-tachogram. The second set of indices contains statistical evaluations for the sequence of the differences of the R-R intervals between subsequent heart beats, i.e.:

$dt_1 = t_1-t_2$ (difference between the first and second R-R interval);
$dt_2 = t_2-t_3$ (difference between the second and third R-R interval);
$dt_3 = t_3-t_4$ (difference between the third and fourth R-R interval), etc.

This sequence is designated as dRR-tachogram to distinguish it from the first.
The following indices are used in the time-domain:

A. From the RR-tachogram
RRA or *NNA* (average value) $=1/_n*(t_1+t_2+...+t_n)$;
SDRR or *SDNN* (standard deviation)

$$=\sqrt{((t_1 - RRA)^2 + (t_2 - RRA)^2 +...+ (t_n - RRA)^2)/_{(n-1)}}\ ;$$

MDRR or *MDNN* (mean (absolute) deviation) $=1/_n*(|t_1-RRA|+|t_2-RRA|+...+|t_n-RRA|)$;

MED (median) – 50% of all values are above and 50% below this value. In a symmetrical distribution this value equals average value. If the two values do not correspond to each other, the distribution of the R-R intervals is not symmetrical.

B. From the dRR-tachogram:
dRRA or *dNNA* (average value) = $1/_{n-1} * (dt_1+dt_2+\ldots+dt_{n-1})$;
dSDRR or *dSDNN* (standard deviation) =

$$\sqrt{((dt_1 - dRRA)^2 + (dt_2 - dRRA)^2 + \ldots + (dt_{n-1} - dRRA)^2)/_{(n-2)}} \, ;$$

dMDRR or *dMDNN* (mean (absolute) deviation) =
$1/_{n-1} * (|dt_1-dRRA|+|dt_2-dRRA|+\ldots+|dt_{n-1}-dRRA|)$;

NN50 (number of pairs of adjacent RR (NN) intervals differing by more than 50 ms in the entire recording);

PNN50 (NN50 divided by the total number of all NN intervals (indicated in %) or number of NN intervals differing by more than 50 ms per minute);

RMSSD or *RMS* (root mean-square deviation of the differences between adjacent RR intervals) = $\sqrt{(1/(n-1))\sum_{1}^{n-1} dt_i^2}$.

As we are not using 24-hour recordings in our studies to determine HRV, *SDARR* is not included in the set of secondary indices from the time-domain.

In the *frequency-domain* the indices result from the spectral analysis of the sequences of R-R intervals through the Fourier transformation.

Signals in the frequency and time range. A cosine signal in the time-domain is characterized by its amplitude A_1 and time interval T_1 between two maximal values (period of oscillation T_1) (Fig 3.2a). The signal is presented both in the time-domain and in the frequency-domain. The frequency f is defined as indirectly proportional to the time t: f=1/t.

The cosine signal in the frequency-domain is therefore presented as a peak with amplitude A_1 and frequency $1/T_1$ (Figure 3.2b). The cosine signal with a lower amplitude A_2 and a shorter period of oscillation T_2 (Figure 3.3a) is shown in the frequency-domain as a peak with frequency $1/T_2$ ($A_1>A_2$, $T_1>T_2$ and $1/T_1<1/T_2$) (Figure 3.3b).

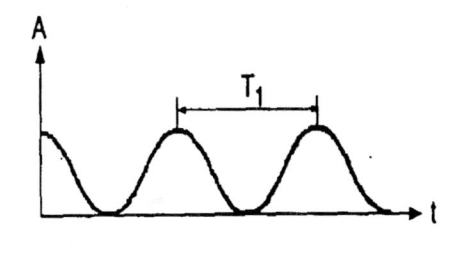

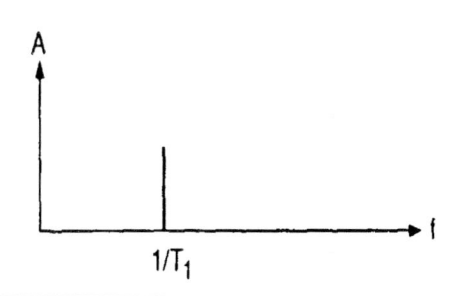

Figure 3.2. Cosine signal in: a). the time- and b). the frequency-domain.

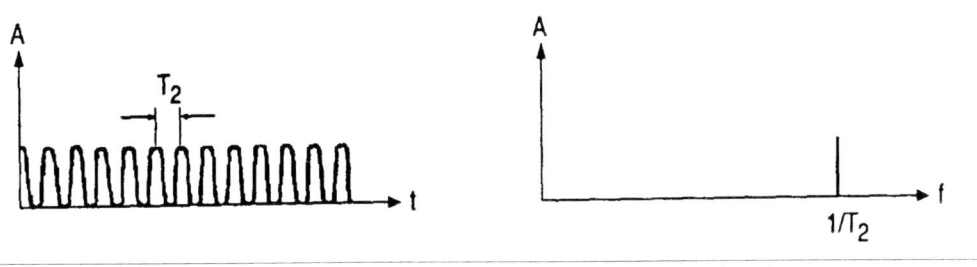

Figure 3.3. Cosine signal in: a). the time- and b). the frequency-domain.

When both signals are summed (Figure 3.4a), two spectral lines result at the points $1/T_1$ and $1/T_2$ (Figure 3.4b).

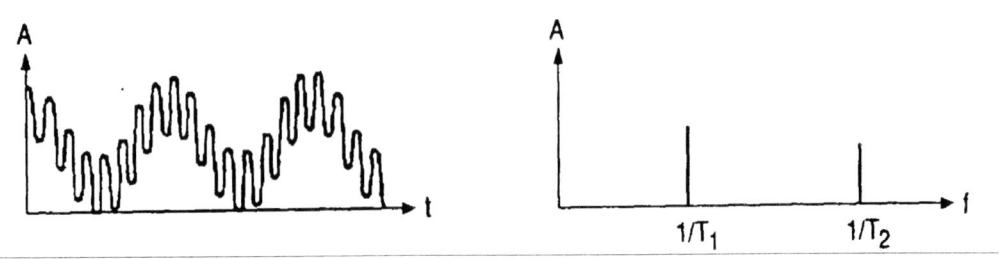

Figure 3.4. Sum of the signals in: a). the time- and b). the frequency-domain.

The Fourier Tansformation

Every arbitrary signal may be presented both within the time and within the frequency-domain, i.e., as a sum of various cosine functions with different amplitudes (Figure 3.5a/b). A transformation from time to frequency-domain is performed by means of the so-called Fourier transformation. A fast type of the Fourier transformation is the so-called Fast Fourier transformation (FFT), which calculates a Fourier transformation with a number of sample values corresponding to a power of 2. In order to perform FFT, frequency analysis examines, for example, $512=2^9$ R-R intervals.

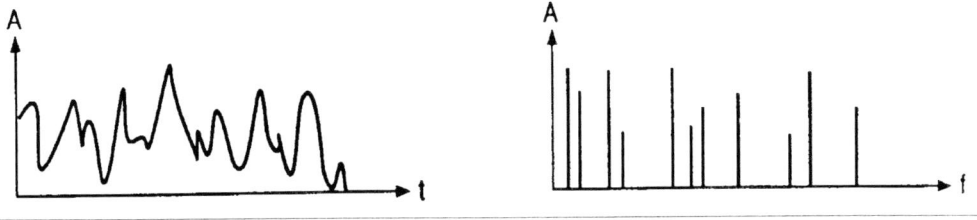

Figure 3.5. Arbitrary signal in: a). the time- and b). the frequency-domain.

Performance of the Fourier Tansformation

Some signal characteristics are easier to detect in frequency in the time-domain. Examinations on oscillation characteristics of motors are performed, e.g., with the Fourier transformation, as indications of wear can be detected at an early stage from changes in the Fourier spectrum.

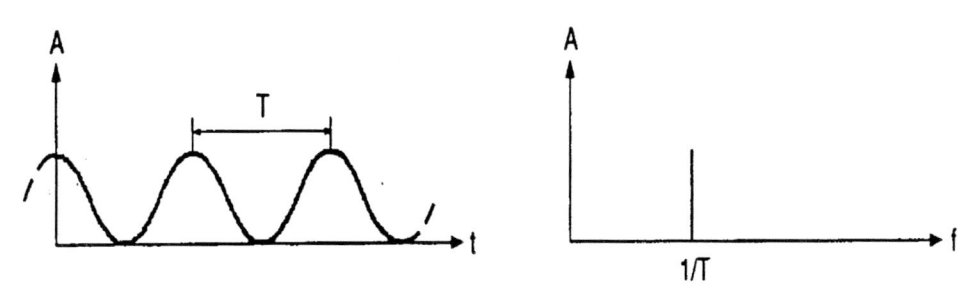

Figure 3.6. Endless periodical signal in: a). the time- and   b). the frequency-domain.

In practice, no endless periodical signals are examined (Figure 3.6a/b), but time-limited non-periodical signals (Figure 3.7a). The Fourier transformation of these time-limited signals does not provide individual peaks as described above, but blurred peaks decreasing in an oscillatory manner to the right and to the left, producing the so-called side-lobes (Figure 3.7b).

Time limitation of Sgnals (windowing)

The usual time limitation (i.e., full signal amplitude up to the point of limitation, otherwise signal amplitude equals zero) can be imagined as windowing of an endless signal (Figure 3.8a) with a rectangular function (Figure 3.8b), speaking of a rectangular window function. Such a rectangular function, however, produces side-lobes with relatively high amplitudes (Figure 3.9b).

Figure 3.7. Time limited signal in: a). the time-   and   b). the frequency-domain.

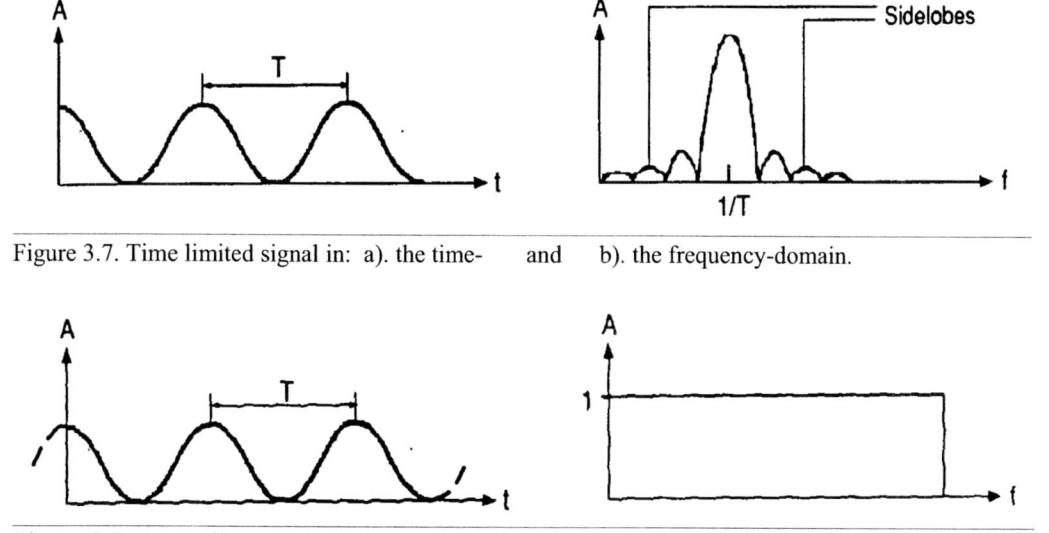

Figure 3.8. An endless periodical signal: a). superimposed with b). a rectangular window.

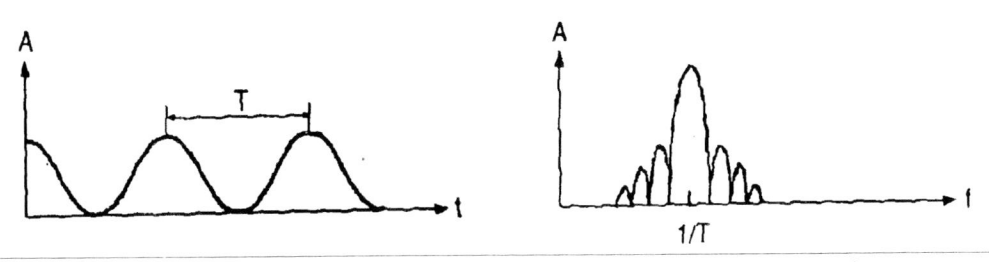

Figure 3.9. Results in a time limited signal in: a). the time- and b). the requency-domain.

Optimal window types

Cutting off the signal not abruptly, but gradually, e.g., by windowing the signal with a cosine-like function (Figure 3.10b), provides side-lobes with lower amplitudes, but extended main lobes (Figure 3.11b), compared to the rectangular window type. Optimal window types, minimizing the disadvantages of high side-lobes and extended main lobes, are the cosine-like Hanning and Blackman-Harris window types.

Standardization

Before performing the Fourier transformation of the R-R intervals, they are standardized according to different procedures. For:

$x(i)$ = value x, calculated from the i-th R-R interval,
$RR(i)$ = i-th R-R interval,
RRA = average R-R interval,
$(1/RR)A$ = average calculated from the reciprocal values 1/RR of R-R intervals,
The R-R intervals are transformed as follows:

1. $x(i) = (RR(i) - RRA)/RRA$;
2. $x(i) = RR(i) - RRA$;
3. $x(i) = (1/RR(i) - (1/RR)A)/(1/RR)A$;
4. $x(i) = 1/RR(i) - (1/RR)A$.

The Fourier transformation is finally applied to the values $x(i)$, and their FFT spectra are analyzed within various frequency ranges.

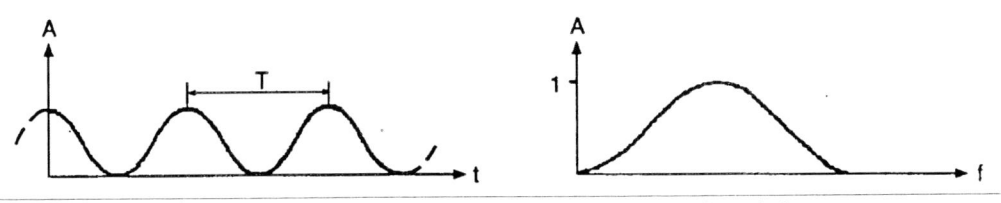

Figure 3.10. An endless periodical signal: a). superimposed with b). a cosine window.

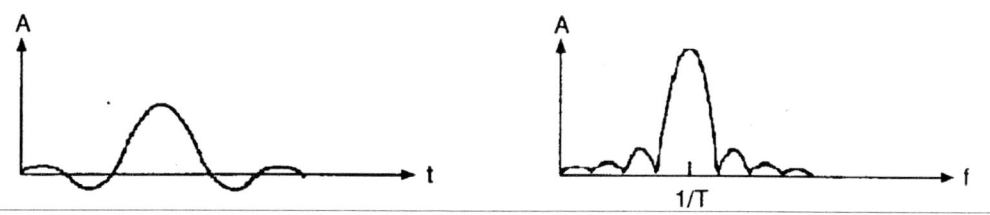

Figure 3.11. Results in a time limited signal in: a). the time- and b). the frequency-domain.

Selectable Parameters before the Analysis

The indicated possibilities for obtaining spectral characteristics of the sequences of R-R intervals lead to different initial conditions for performing FFT. For comparability between the results of the spectral analysis of HRV it is necessary to indicate specifically the chosen

- FFT window type;
- number of R-R intervals to be measured (128, 512, etc);
- type of preparation and standardization of R-R intervals before FFT is applied.

Frequency Spectrum and Analyzed Ranges

After frequency analysis (with FFT), the distribution of the R-R intervals is presented within the frequency-domain. The spectral high-frequency limit is 1/(2RRA). With short recordings (below 10 minutes), the analyzed frequency ranges are differentiated as very low frequency (VLF), low frequency (LF) and high frequency (HF) (in ms^2), and as LF/HF ratio.

With a certain difference in the cited limits of these frequency ranges, we used in our studies for VLF 0.016 – 0.05Hz; for LF 0.05 – 0.15Hz; for HF 0.15 – 0.35 Hz.

PHYSIOLOGICAL MECHANISMS DETERMINING HRV

In norm, the variation in the duration of the R-R intervals is provoked by different *periodic and non-periodic* impacts on the cardiac pacemaker.

The non-periodic impacts are connected with arbitrary events, hence they are not of interest from the viewpoint of the physiological mechanisms determining HRV. The algorithms for analysis of the variability comprise procedures or prescriptions for eliminating the effects of arbitrary impacts, because they could have a substantial influence on the results of the analysis (the so-called editing of the sequences of R-R intervals).

The phases of respiration have a periodic impact: elevation of the heart rate during inhaling and its reduction during exhaling. This periodic fluctuation in ECG is known as respiratory sinus arrhythmia (RSA). RSA is provoked above all by the fact that the vagal efferent traffic towards the sinus node is synphasic during exhaling, and is lacking or weak during inhaling. RSA prevails during rest. The non-respiratory sinus arrhythmia is associated with the slower-wave modulation of the heart rate, whereby the impact is achieved along the sympathetic and humoral pathways. It is more tangibly manifested under different loading of the organism.

The attraction of HRV as a method for clinical diagnostics, control on the quality of the treatment and prediction of the course of the disease in cardiology is determined by the possibility to link the variability and the state of the cardiovascular system within the frameworks of a simple model [2] (Figure 3.12). The rhythmic nature of the cardiac activity is determined both by the automatism of the sinus node and by the specificities of the conducting system of the heart, and also by the stimulating or inhibiting innervation of these structures by the autonomic nervous system. The sinus and the atrioventricular nodes are richly innervated by the parasympathetic nerve fibers. The higher activity of these nerves leads to lowering of the heart rate. Sympathetic innervation comprises all structures of the heart: the pacemaker, the conduction system, the atrial and the ventricular muscles. The growing sympathetic activity increases the heart rate. The balance between the activities of the two parts of ANS at any moment reflects its regulating function on the cardiac activity with a view to maintaining such a blood circulation that is optimal for the adaptive response of the organism to functional changes (see Chapters One and Two). The feedback is achieved through the baroreflex mechanism, which is activated by the baroreceptors in most of the major arteries.

Obviously, the adoption of the cited model is connected with the notion of the extent to which the HRV indices reflect the activity of the two parts of the ANS and the autonomic balance (AB). Many materials have been published on this issue. It is interesting to note the discussion in *Circulation* [3] in connection with Eckberg's critical review entitled: Sympathovagal balance: a critical appraisal. Eckberg disputes the dominating views that:

- in the frequency band of up to 0.1 Hz (predominantly LF) the spectrum of the R-R intervals is mediated by fluctuations in the activity of the sympathetic nerve;
- the frequency band HF is mediated almost exclusively by the activity of the vagal nerve;
- the physiological interventions provoke a reciprocal change of the sympathetic and vagal response, i.e., that the autonomic balance is a ratio of neural mechanisms with opposite responses.

Eckberg's claims are refuted in three commentaries: of Sleight and Bernardi [4], of Malliani, Pagani, Montano and Mela [5], and of Malek [6]. This is particularly important because the discussion preceded the publication of the *Task Force...* and the physiological correlates of HRV cited there have a valid argumentation. The opponents believe that Eckberg's conclusion result from misconception of the methodologies used in the publications cited by him.

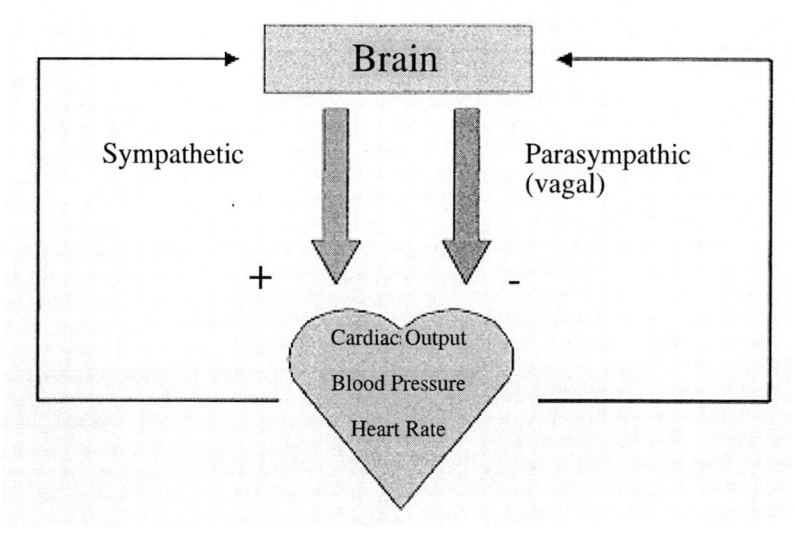

Figure 3.12. Schematic working model.

Briefly, the opponents claim that:

- the low-frequency range is defined both by sympathetic and by vagal influences, with domination of the sympathetic component;
- the respiratory fluctuations of the cardiac rhythm are clearly distinguished from the LF interval;
- the sympathetic-vagal (the autonomic) balance is formed by a reciprocal functional link between the two types of cardiac innervation.

We accept these views because, in our opinion, they correlate well with the idea about a common origin of the changes in the heart rate and HRV. The interval between the cardiac cycles depends on the rhythmic activity of the pacemaker cells of the sinus node. This rhythmic activity is under constant nerve and endocrine control and under the influence of a number of humoral factors that change the threshold of the spontaneous depolarization of the pacemakers of the sinus node. This leads to increase – or decrease accordingly – of the interval between the cardiac cycles and hence of the heart rate. Therefore, the factors regulating the heart rate will also determine the HRV. An important characteristic of this process is the determined periodicity (including circadian nature), with which the activity (the level of the impact) of the indicated factors changes.

It can be considered to be proven that the range of HF in HRV is connected with respiration. The respiratory genesis of HF is confirmed by the frequently observed coincidence between the respiration rate and the frequency of the high-frequency peak in the spectrum. With simultaneous registration of the respiration and ECG it can be seen that in most cases the R-R interval diminishes with each inhaling, and increases with each exhaling [7]. At the same time, the mechanism of the interconnection between the respiratory cycles and the duration of the R-R interval has not been completely clarified. At least four hypotheses concerning the respiratory modulation of the heart rate are known [8], [9,10,11], [12,13], [14,15]. However, it has been determined with certainty that the vagus is an efferent link in that mechanism. The respiratory modulation of the heart rate disappears, for example, when the vagus is severed [16]. The physiological determinateness of the HF components in HRV, predominantly of the parasympathetic negative chronotropic influences on the pacemakers of the sinus node, has been proven experimentally in [16,17]. Obviously, the high-frequency components of HRV are determined in the long run by the link between the vagus and the sinus node. *This gives grounds to judge from the spectral power in the HF area about the state of the parasympathetic part of ANS.*

The LF spectral area of HRV comprises frequencies predominantly from 0.05 to 0.15 Hz, with frequent presence of dominating spectral component of 0.1 Hz. A periodic component with such frequency exists in the arterial pressure as well. Using cross-spectral analysis of the fluctuations in the arterial pressure and the heart rate, it has been found that the prolongation of the R-R intervals is preceded by elevation of the arterial pressure [18]. This is accepted as evidence that the 0.1 Hz component in LF is due to the respective periodic component in the arterial pressure and this effect is explained with the baroreflex mechanism. This means that the sympathetic activity is suppressed in response to the elevated AP and the parasympathetic activity increases, which leads to prolongation of the R-R intervals. Many researchers accept that the 0.1 Hz component in AP results from generalized bursts of sympathetic vasomotor activity with the same frequency [19,20,21,22]. Other authors believe that the rhythm of the generalized sympathetic activity is induced by the baroreceptor structures as indicated in [18,23]. Upon lowering of AP below a certain level, the baroreceptors are activated, which causes an increase in the sympathetic vasomotor activity and narrowing of the vessels. As a result, the AP rises, reaches a maximum and begins to drop. The cycle is repeated many times. There exist two other hypotheses as well: on the generating of a 0.1 Hz rhythm in LF as a result of the rhythmic character of the myogenic responses of the arterioles, which changes the HR through the baroreflex mechanism [24], and on the inducing of generalized sympathetic activity by an oscillator placed in the brain stem [9,25]. Precisely this oscillator determines the periodic changes in the intensity of the impulses from the sympathetic vasomotor neurons. These oscillations along the sympathetic efferent nerve fibers are transmitted to the heart and to the vessels, and they induce innervation of the cardiac sympathetic structures responsible for the basal innervation of the organ.

Even if we assume that the three mechanisms (baroreflex, central and myogenic) participate in the formation of the LF range of HRV, it is important from a practical point of view for it to be connected with the activity of the postganglionic sympathetic fibers, and it is possible to judge from its spectral power about the state of the sympathetic control of the heart rate. Nevertheless, frequency subintervals with genesis from the vagal activity have been identified in the range of LF [1]. This prompts us to accept that *the integral power of LF reflects both sympathetic and parasympathetic influences on the sinus node, with prevalence of the sympathetic ones.*

In the cited models for the formation of the HF and LF ranges in HRV, it is logical for the LF/HF ratio to reflect the balance between the sympathetic and the parasympathetic activity of the VNS, i.e., of the autonomic balance. Certain facts are in favor of this assumption. In upright position of young healthy individuals there is a considerable increase of the integral spectral power in the LF range. At the same time, a more pronounced periodicity of the sympathetic impulsation is registered, with simultaneous rise of the amplitude of the impulses [26]. The amplitude of the high-frequency components during transition from lying to upright position decreases, even to the extent of total disappearance. Consequently, it is possible to speak about reciprocal ratio between the low-frequency and the high-frequency components in the HRV spectrum. A similar correlation is also observed between the activity of the sympathetic and of the parasympathetic part of the vegetative control of the heart [1,27].

The mechanisms for formation of the VLF range in the HRV spectrum have not been completely clarified. The wave activity corresponding to this frequency interval has a period exceeding 25 ms. Most probably, VLF is formed under the influence of the rhythmic character of the thermal regulation by the hypothalamus [28] and of hormonal changes. The link between the VLF reduction and the activation of the renin-angiotensin system has been proven [29]. Other authors [30] indicate that angiotensin inhibits the parasympathetic activity through central neural mechanisms. Petretta, Bianchi and Marciano [31] have shown that patients with left-ventricular hypertrophy demonstrate a smaller increase of VLF during the night, and they assume that this is due to the activation of the renin-angiotensin system in hypertensive patients during the night and the loss of the normal circadian rhythm of the hormonal secretion. In spite of the different views on the formation of the VLF range in the HRV spectrum, most authors assume that it *characterizes the activity of the sympathetic part of the VNS.*

Correlations between frequency- and time-domain indices. Special attention is devoted both in the *Task Force...* and in other materials summarizing the methods for HRV evaluation on the correlation in principle between the indices from the *frequency- and time-domains*. The correlation coefficient exceeds 0.80 for some pairs of indices [32]. The following correlations are characteristic of the most frequently used indices from both areas:

- SDRR and total power (TP) (r=0,85): the increase of their values is connected with higher vagal activity, the decrease – with enhancing of the sympathetic activity and inhibition of the vagus;
- RMSSD; PNN50 and HF (r=0,87; r=0,88): indices for evaluation of the activity of the parasympathetic part in AB.

These tendencies of unidirectional change in the respective indices from the time- and frequency-domains are particularly important for the studies conducted by us and summarized in Chapters 5, 6, 7 and 8, because their presence or absence attributes reliability or conditionality of the results obtained.

STANDARDS OF THE MEASUREMENTS

In our studies, discussed in Chapters 5, 6, 7 and 8, we have adopted standards for measurements and evaluation of HRV, which correspond to the recommendations contained in the *Task Force...*, on the one hand, and on the other – to the specificities of the method proposed by us for evaluation of time-related changes in HAB (see Chapter 5).

We used recordings of R-R intervals during rest and upon VNS stimulation with the handgrip test and Valsalva manoeuvre. For every individual the HRV indices were measured on 512 RR intervals during rest, 128 RR intervals with the handgrip test and 64 RR intervals with the Valsalva manoeuvre. The R-R intervals for the three tests were registered in the morning (8-9 a.m.) and in the afternoon (2-3 p.m.). Before the ECG registration, the persons were kept lying for 30 minutes at room temperature of 22-24 °C. The tests were conducted in lying position, with 20-minute interval between the tests. The ECG signals were sampled with 1000 Hz and with 2.5 microV digital resolution (dynamic range: ± 10 mV AC).

From the indices indicated in 3.1. HRV we used SDRR, MDRR, dSDRR, dMDRR, PNN50, RMSSD, VLF, LF, HF and LF/HF. This set, as was pointed out already, has a well clarified physiological correlation with the changes in the two parts of VNS, and it is also most discussed in the literature. We believe hence that comparability is thus secured between the results obtained by us and the results of other researchers. We used all HRV indices indicated in 3.1 only due to the specificity of the study in section 6.6.

The R-R intervals before the spectral analysis were normalized with respect to the mean value. We used the Hanning spectral window type. It was not necessary to normalize LF and HF compared to the total power because the changes in AB were determined by comparing values for the same HRV indices in two time intervals. The duration of the measurement with the Valsalva manoeuvre does not allow the determination of the VLF index. The brief duration of the recordings used and the indicated conditions for recording the ECG-signals from the subjects studied allow us to consider that the problems connected with non-stationarity have been evaded.

REFERENCES

[1] Task Force of the European Society of Cardiology and the North American Society of Pacing and Electrophysiology. Heart rate variability. Standards of measurement, physiological interpretation and clinical use. *Circulation*, 1996, 93, 1043-1065.

[2] Aubert, AE; Ramaekers, D; Beckers, Г; Leuven, KU. *Heart Rate Variability*. Reprint from HF-Cont@kt, issue 4, 2002.

[3] Eckberg, DL. Sympathovagal balance: a critical appraisal. *Circulation*, 1997, 96, 3224-3232.

[4] Sleight, P; Bernardi, L. Sympathovagal balance. *Circulation*, 1998, 98, 2640.

[5] Malliani, A; Pagani, M; Montano, N; Mela GS. Sympathovagal balance: a reappraisal. *Circulation*, 1998, 98, 2640-2643.

[6] Malik, M. Sympathovagal balance: a critical appraisal. *Circulation*, 1998, 98, 2643-2644.

[7] Kotelnikov, SA; Nozdratchev, AD; Odinak, MM; Shustov, EB; Kovalenko, IY; Davidenko, VY. Heart rate variability: a notion of mechanisms. (In Russian). 2003:01:05. Availiable from: URL: Dr. Med. Ru. *www. Medlincs.ru/article. php?sid=7234.*

[8] Nozdratchev, AD. Axon-reflex. A new opinion in old area. (In Russian). *Physiol J*, 1995, 81, 136-144.

[9] Anrep, GV; Pascual, W; Rossler, R. Respiratory variation of the heart rate. The reflex mechanism of the respiratory arrhythmia. *Biol Sci*, 1936, 119, 191-217.

[10] Richter, DV; Spyer, KM. *Cardiorespiratory control. General regulation of autonomic function.* New York: Oxford Univ Press, 1990, 189-207.

[11] Montano, N; Gnecchi Ruscone, T; Porta, A; et al. Presence of vasomotor and respiratory rhythms in the discharge of single medullary neurons involved in the regulation of cardiovascular system. *J Auton Nerv Syst*, 1996, 57, 116-122.

[12] Melcher, A. Carotid baroreflex heart rate control during the active and assisted breathing cycle in man. *Acta Physiol Scand*, 1980, 108, 165-171.
[13] Akselrod, S. *Components of heart rate variability. Heart rate variability*. New York: Armonk, 1995, 146-164.
[14] Lucy, SD; Hughson, RL; Kowalchuk, JM; et al. Body position and cardiac dynamic and chronotropic responses to steady-state isocapnic hypoxaemia in humans. *Exp Physiol*, 2000, 85, 227-237.
[15] Al-Ani, M; Forcins, AS; Townend, JN; Coote, JH. Respiratory sinus arrhythmia and central respiratory drive in humans. *Clin Sci*, 1996, 90, 235-341.
[16] Rimoldi, O; Pierini, S; Ferrary, A; et al. Analysis of shot-term oscillations of R-R and arterial pressure in conscious dogs. *Am J Physiol*, 1990, 258, 967-976.
[17] Chess, GF; Tam, RM; Carlaresu, FR. Influence of cardiac neural inputs on rhythmic variation of heart period in cat. *Am J Physiol*, 1975, 228, 775-780.
[18] Karemaker, JM. Analysis of blood pressure and heart rate variability: theoretical consideration and clinical applicability. In: Low PA, editor. *Clinical autonomic disorders. Evaluation and management*. Boston: Little Brown and Co; 1993; 315-330.
[19] Press, G; Polosa, C. Patterns of sympathetic neuron activity associated with Mayer waves. *Am J Physiol*, 1974, 226, 724-730.
[20] Pagani, M; Lombardi, E; Guzzeti, S; et al. Power spectral analysis of heart rate and arterial pressure variabilities as a marker sympatho–vagal interaction in man and conscious dog. *Circ Res*, 1986, 59, 178-193.
[21] Lombardi, E; Montano, N; Fnocchiaro, S; et al. Spectral analysis of sympathetic discharge in decerebrate cats. *J Auton Nerv Syst*, 1990, 30, 97-100.
[22] Saul, JP; Rea, RF; Eckberg, DL; et al. Heart rate and muscle sympathetic nerve variability during reflex changes of autonomic activity. *Am J Physiol*, 1990, 258, 713-721.
[23] De Boer, RV; Karemaker, JM; Time delays in the human baroreceptor reflex. *J Auton Nerv Syst*, 1983, 9, 399-409.
[24] Janssen, BJA; Oosting, J; Slaff, DW; et al . Hemodynamic basis of oscillations in systemic arterial pressure in conscious rats. *Am J Physiol*, 1995, 62-71.
[25] Cevese, A; Grasso, R; Poltronieri, R; Schena, F. Vascular resistance and arterial pressure low-frequency oscillation in the anesthetized dog. *Am J Physiol*, 1995, 268, 7-16.
[26] Pagani, M; Montano, N; Porta, A; et al. Relationship between spectral components of cardiovascular variabilitis and direct measures of muscle sympathetic nerve activity in humans. *Circulation*, 1997, 95, 1441-1448.
[27] Montano, N; Ruscone, TG; Porta, A; et al. Power spectrum analysis of heart rate variability to assess the changes in sympatovagal balance during graded orthostatic tilt. *Circulation*, 1994, 90, 1826-1831.
[28] Stauss, HM. Heart rate variability. *Am J Physiol*, 2003, 285, 927-931.
[29] Akselrod, SD; Gordon, D; Ubel, FA; et al. Power spectrum analysis of heart rate fluctuation: a quantitative probe of beat-to- beat cardiovascular control. *Science*, 1981, 213, 220-222.
[30] Lumbers, ER; McCloscey, DI; Potter, EK. Inhibition by angiotensin II of baroreceptor-evoked activity in vagal efferent nerves in dogs. *J Physiol*, 1978, 294, 69-80.

[31] Petretta, M; Bianchi, V; Marciano, F. Influence of left ventricular hypertrophy on heart period variability in patients with hypertension. *J Hypertens*, 1995, 13, 1299-1306.

[32] Bigger, JT; Albrecht, P; Steinman, RC; Rolnitzky, LM; Fleiss, JL; Cohen, RJ. Comparison of time-and frequency-domain-based measures of cardiac parasympathetic activity in Holter recordings after miocardial infarction. *Am J Cardiol*, 1989, 64, 536-538.

Chapter 4

CIRCADIAN CHARACTERISTICS OF THE CARDIOVASCULAR SYSTEM

Rada Prokopova
St. Anna University Hospital, Sofia, Bulgaria

Mikhail Matveev
Centre of Biomedical Engineering, Bulgarian Academy of Sciences, Bulgaria

CIRCADIAN RHYTHMS

Circadian rhythms have been found in animals at every level of eukaryote organization, including in humans. These rhythms are usually associated with the 24-hour circadian cycle. In the past it was believed that circadian rhythms are a *passive* reaction of the organism to the periodic change of external factors. In recent decades it has been proven that the inner rhythms are preserved even when all factors of the environment have been excluded. The periodicity of these "free" rhythms is shorter or longer than 24 hours. This also indicates that rhythmicity is endogenous in nature. This endogenous rhythmicity has received the general name of "biological clock." Since endogenous rhythms correspond only approximately to the circadian cycle, they are referred to as circadian (from the Latin words *circa* – around and *dies* – day). The free circadian rhythms do not attenuate for a long time, i.e., they have the properties of a self-excitatory oscillator. Usually the frequency of the inner rhythms is synchronized with the circadian cycle.

The genes controlling the circadian rhythms (genes: per, frq, clock, tau) have been discovered in some organisms (*Drosophila melanogaster*, *neurospora*, mouse and golden hamster). In 1971, Konopka and Benzer [1] succeeded in identifying the region in the X chromosome of *Drosophila*, which controls the circadian rhythm. Bardgiello, Jackson, Yong [2] have demonstrated that by injection of a fragment of genes responsible for the biological clock (per clock genes), the circadian nature can be restored completely in *Drosophila* mutants with missing circadian rhythm. These data give the first evidence that the biological clock is genetically determined.

The genes determining "the biological clock" are deeply inherent to the evolutionary process. This has been proven through their similarity in *Drosophila*, in mammals and in man [3]. Obviously, the evolution has preserved and is transmitting the molecular base of the circadian rhythms. It is important to know that the endogenous biological rhythms are primary in nature and adapt the living systems better and more easily to the changing environment during the 24-hour cycle and the alternating seasons.

More than 100 different physiological parameters having cyclic fluctuations with a 24-hour period have been identified in man. For example, temperature of the human body is the lowest early in the morning, reaching a maximum in the evening, the fluctuation being between 1 and 1.5 °C.

The most pronounced and the most important circadian rhythm is the sleep-wakefulness cycle, where essential changes occur in the functioning of the organism: lowering of the temperature, the pulse rate, the respiration rate and the arterial pressure are observed during sleep. However, it has been proven in many experiments that these fluctuations of the physiological parameters are also detected in sleep deprivation. Similar experiments have demonstrated presence of numerous circadian oscillators, slightly differing in frequency, in man and in other highly organized creatures. These oscillators are connected with the sleep-wakefulness cycle due to synchronization between them or with external signals. In experiments carried out in special underground chambers or caves it has been proven that in man the circadian rhythms are preserved even in the case of isolation from the environment. In that case the period of the rhythms is a little longer than 24 hours. Similar experiments have revealed that the individual oscillators have different periods and are relatively independent. In the case of total isolation from the environment it has been found, e.g., that in the first 24 hours there is a good relation between the circadian rhythm of the temperature and of the sleep-wakefulness cycle, but that relation is gradually lost. The rhythms of the vegetative functions are desynchronized with respect to the sleep-wakefulness cycle, preserving their own rhythm of 25 hours.

The inner clock of humans and mammals is located in the suprachiasmatic nucleus of the central brain. It consists of around 10,000 neurons, on both sides of the median line above chiasma opticus [3]. If this nucleus is destroyed (in animals – experimentally and in humans as a result of some disease, e.g., tumor compression), the circadian rhythm is disturbed. It has been proven that if neonatal cerebral hypothalamic tissue containing cells from suprachiasmatic nucleus is transplanted to animals with ablation of the suprachiasmatic nucleus, the circadian model "activity – rest" is restored. Obviously, "the biological clock" is a property of the suprachiasmatic nucleus, and the experiment demonstrates the possibility of its restoring through transplantation.

CHRONOBIOLOGY OF THE CARDIOVASCULAR SYSTEM

Ever since the 17th century it has been known that there is a difference in the heart rate and in the arterial pressure during the day and during the night. In the 20th century, with the appearance of the systems for 24-hour monitoring of the heart rate and arterial pressure, the circadian characteristic of these parameters for the activity of the cardiovascular system in norm and in pathology was identified.

Experiments were carried out to determine the possible circadian character of hormonal secretion. The link between the 24-hour hormonal levels – especially of the catecholamine, cortizol, renin-angiotensin-aldosterone secretion – and the changes in the cardiovascular system is the subject of numerous studies. In healthy subjects it has been found that the average levels of catecholamines (adrenaline and noradrenalin) are higher during the day compared to those during the night. The difference with the two catecholamines consists in the fact that the adrenaline level begins to rise immediately after waking, while the elevation of the noradrenalin level starts around four hours after waking. The peak in the levels of both hormones is attained around 11 a.m. [4].

The secretion of cortizol also has a circadian characteristics, similar to the catecholamines secretion, but with an earlier morning peak. The lowest values are measured at 9 p.m., after 3 a.m. a rise in the secretion starts, reaching its maximum at 7 a.m. [4].

It is known that the secretion of renin [5], angiotensin and aldosterone [6] is also higher in the morning hours.

Healthy individuals reveal a circadian nature in the coagulation and fibrinolysis as well. Hypercoagulation is observed in the morning hours, whereby the susceptibility to platelet aggregation increases [6]. Platelet aggregation induced by adrenaline and adenosine diphosphate (ADP) is increased only after getting up in the morning; there is no increase in aggregation after 12:30 p.m. In an experiment in which the patients remain lying in bed, there is no rise in the platelet aggregation. Such a difference is explained with the increased hormonal secretion (catecholamines and R-A-A) upon standing up compared to lying in bed [6].

Circadian variability is also observed in the fibrinogen level: the level is the highest in the morning, which is also associated with a rise in the cortizol secretion. Blood viscosity is also higher [7].

Similar circadian changes have been discovered for fibrinolysis. The plasminogen activator inhibitor (PAI) and tissue plasminogen activator (t-PA) are key components of fibrinolysis. t-PA increases fibrinolysis, while PAI inhibits it by binding to t-PA. the plasma PAI activity is with peak at 6 a.m. The plasma activity of the t-PA-antigen level is with the same peak at 6 a.m. This also determined the lowest level of fibrinolytic activity in the morning [8].

CIRCADIAN ACTIVITY OF THE AUTONOMIC NERVOUS SYSTEM

The circadian characteristic of the arterial pressure and the heart rate is a reflection of *the circadian activity of the autonomic nervous system*. As was indicated already, the variability of the heart rate is an indicator of the activity of the autonomic nervous system and it can be used for evaluation of the autonomic balance. Numerous experiments have been made to determine the changes in the autonomic balance in the different hours of the 24-hour cycle. Massin, Maeyns, Withofs [9] studied 57 healthy children aged from 2 months to 15 years. The study also included 5 children with diabetes, without clinically perceptible vagal neuropathy. They revealed progressive maturing of the activity of the autonomic nervous system. Circadian variation in the heart rate exists since the age of 4 months, but circadian nature in most of the indicators for HRV is found after the age of 12 months. This is probably due to the fact that after this age the individuals already have a normal sleep/wakefulness cycle. The opinion is imposed that the reason for the circadian change in ANS, and hence in HRV as well, is the actual sleep state and the "maturing" of the biological clock. The biological clock has an important role for determining the wakefulness/sleep intervals and for the organization of sleep, as well as for the coordination of sleep with other physiological rhythms, notably those of the autonomic balance [9].

In healthy individuals, the reason for the circadian nature seems to be the lowering of the sympathetic tone rather than the change in the tone of the parasympathicus. This assertion is supported by the existence of circadian nature in diabetes patients as well. Malpas and Purdie [10] have found circadian changes in HRV in 23 adults (healthy and diabetics) with increase in the variability during the night, whereby the change does not depend on the vagal neuropathy in the diabetics. In view of this, the authors conclude that the circadian characteristic of the autonomic balance is due to fluctuations in the activity of the sympathetic tone. The lowered sympathetic tone during the night is associated with the reduction of the circulating catecholamines, but it is also demonstrated and through stimulation of the sympathetic nerve, whereby a slower response is registered.

Other authors [11] have studied the circadian characteristic of HRV in healthy individuals aged from 20 to 70 years. According to them, ageing causes reduction of the nocturnal vagal tone.

There are also studies on the autonomic balance in both genders. The data are based on the frequency analysis of HRV in 105 healthy individuals aged from 20 to 78 years [12]. The LF components were found to be higher during the day, and HF – during the night, in all age groups, but among the young these circadian variations are greater and decrease with age. The study involving 15 healthy individuals aged 20-39 years and 40 aged 60-74 years revealed that the amplitude of the circadian changes in the activity of the autonomic nervous system decreases with the years, even in 7% of the older individuals there exists an inversion in the activity [13].

The young men have high LF values during the day, while the HF values women remain high during the day irrespective of their aged [12,13]. The result points to a certain susceptibility to hypersympathicotony among the men and maybe a higher vagal tone in women during the day. These finds can be associated with the different cardiovascular risk in both genders.

CIRCADIAN NATURE IN THE AUTONOMIC BALANCE IN CARDIOVASCULAR DISEASES

Circadian Characteristic of Arterial Hypertension

It has been known for decades that arterial pressure has a circadian characteristic. After the introduction of systems for 24-hour monitoring of the arterial pressure, a different circadian profile of the arterial pressure was found in hypertensive individuals. Some preserve the normal characteristic with a fall of the arterial pressure during the night (dippers). In others this fall is absent (nondippers). A dependence is found between the absence of nocturnal fall and some types of secondary hypertension: in diabetes, renal failure, pheochromocytoma, hypercorticism, as well as in the case of development of left-ventricular hypertrophy and cerebrovascular disease. The absence of nocturnal fall is also associated with a higher probability of appearance of cardiovascular incidents. The plasma catecholamine levels decrease during sleep and play a major role for the existence of nocturnal fall [14,15].

The link between hypertension and hypersympathicotony is known. It is associated, on the one hand, with pre-excitation of the cerebral centers, and on the other – with elevated catecholamines level. Many studies are devoted to the dependence between the appearance and the progression of hypertension and the disordered autonomic balance. In the case of borderline hypertension, Anderson, Sinkey, Lawton [16] have studied using microneurography the muscle sympathetic nerve activity, the plasma levels of catecholamines and have proven the existence of hypersympathicotony. It leads to hyperkinetic syndrome with higher heart rate, increased cardiac output and increased vascular tone. On the other hand, the increased sympathetic tone is connected with the development of left-ventricular hypertrophy and hypertrophy of the intima of the vessels and indirectly – through activation of the RAAS. Most researchers find that the circadian profile specific for healthy individuals, with prevalence of the sympathicus during the day and of the parasympathicus during the night, is preserved in the case of hypertension as well, without development of left-ventricular hypertrophy [17]. Other authors [18] find a disordered circadian profile with higher LF values, indicating high tone of the sympathicus during the night. Muiesan, Rizzoni, Zuli [19] have found that, compared to normotensive individuals, hypertensive individuals without left-ventricular hypertrophy have significantly higher LF components during the day, but not during the night.

The progressing of hypertension changes also the circadian characteristic of HRV, which is connected with a change in the autonomic balance. In patients with left-ventricular hypertrophy and missing nocturnal fall, the plasma level of catecholamines is higher than that of hypertensive individuals with preserved circadian profile of AP [17]. Gusetti, Dassi, Pecis [18] find a connection between the progression of the hypertensive disease and the reduction of the circadian difference between the LF component in the spectrum. Probably the progression of hypertension damages the circadian nature in the activity of the sympathetic tone. Cakko, Mulingtapang, Huikury [20] also have a similar find. They have found that circadian rhythm of LF is absent in hypertensive individuals with left-ventricular hypertrophy, while the power of HF, which reflects the activity of the parasympathetic part, is lower among the hypertensive individuals compared to the controls. The reduction of the physiological circadian fluctuation of the autonomic nervous system is also confirmed by Kohara, Nishida, Maguchi [17]. They find a significant reduction of HF during the night in hypertensive patients with absence of nocturnal fall, which presupposes lowering of the parasympathetic tone. Muiesan, Rizzoni, Zuli [19] have found in hypertensive individuals with left-ventricular hypertrophy an increase of the LF component both during the day and during the night, and reduction of the HF component. It is assumed that this difference is due to hypersympathicotony or reduction of the tone of the parasympathicus in hypertensive individuals with left-ventricular hypertrophy. Almost all studies point also to a general reduction of HRV in hypertensive individuals with left-ventricular hypertrophy, whereby the values for the day are lower than those for the night.

Summarizing these studies, it can be said that the appearance and progression of the hypertensive disease is connected with damaged circadian characteristic of the autonomic balance, with manifestations of hypersympathicotony or reduced parasympathetic tone.

Circadian Characteristic of the Myocardial Ischaemia

The development of 24-hour monitoring made it possible to detect transient myocardial ischaemia. It was found that 70%-80% of the ischaemic episodes are silent myocardial ischaemia. The peak of the ischaemic episodes is between 6 a.m. and noon, with double increase in the first two hours after getting up [21], followed by a smaller increase in the evening peak. This circadian model of distribution of the ischaemic episodes exists in the patients both with stable and with unstable angina.

In the variant angina, a peak was found in the elevation of the ST-segment and anginous symptoms during the treadmill test in the morning hours. In the case of stable angina, the frequency of symptomatic and asymptomatic ischaemic episodes, identified through depression of the ST-segment in outpatient ECG studies, is higher in the morning hours, being strongly influenced by physical activity in the morning [22].

Circadian Characteristic of Myocardial Infarction

Already back in 1910, Obraztsov and Strazhesko [e.g. in [23] assumed that the incidence of myocardial infarction is higher in the morning hours due to the start of the daily activity. Their assumption found confirmation with the development of research methods. The MILIS study (Multicenter Investigation of Limitation of Infarction Size) proved by studying the creatin phosphokinase that the incidence of infarctions is much higher between 6 a.m. and noon [24]. This find is confirmed by other studies as well, in which pain is a marker for the hour of the onset of the ischaemic incident. The characteristic peak of infarctions in the morning is detected in young individuals and adults, in men and women, in smokers and non-smokers, in people who drink coffee and others who do not, in individuals with prior infarction and without such incident.

Many endogenous rhythms have circadian characteristics similar of those of infarction and are connected with it. It was mentioned already that the cortizol level has a peak around 6 a.m. and then it gradually decreases. Catecholamine levels start rising around 7 a.m. and reach their maximum around 11 a.m. The heart rate is the highest again during these hours. The increased activity of the sympathetic nervous system in the morning induces general and coronary vasoconstriction, a rise in the arterial pressure and probability of rupture of the intima of the plaque. These circadian changes are combined with increased susceptibility to thromboses and reduced fibrinolytic activity in the morning hours, and probably this constellation of factors is the reason for the peak in the frequency of the ischaemic incidents in the morning [25].

The ISAM study – Intravenous Streptokinase in Acute Myocardial Infarction [26] - gives a typical circadian characteristic in the frequency of the onset of myocardial infarction: it is pointed out that the frequency is 3.8 times higher between 8 and 9 a.m. (peak of the incidents) than between noon and midnight. This find is confirmed with studies of creatinphosphokinase and with the hour of onset of the pain. It is important to note that the circadian model is stable in individuals with low and high ejection fraction, with low and high index of dyssynergy, young and adults, women and men, with, one- two- or three- vessel disease, in the groups with stenoses below 75% and above 95%, in patients with anterior and posterior infarction, in trained and untrained with streptokinase. Only one subgroup demonstrates absence of increased frequency of MI in the morning: the patients treated with beta-blockers. A probable reason for this is the reduction of hypersympathicotony in the morning hours. Other researchers associate the elevation of the MI incidence in the first three hours after getting up from sleep in the morning with the increased platelet aggregation cited in other studies as well. The Physicians Health Study [27] investigated 342 cases of non-fatal MI, in 211 of which there is information about the hour of onset of the infarction. The distribution of the hours of onset of the non-fatal MI has a maximum between 4 and 10 a.m., and a second smaller peak late in the evening. This circadian nature is observed mainly in the placebo group, while the morning peak is considerably lower in the group receiving aspirin. The fact concerning the onset of MI in the first hours after getting up from sleep in the morning is also confirmed by Goldberg, Brady, Muller [28]. This is probably due to the numerous changes in the physiological processes in the morning after getting up from sleep. Hypersympathicotony, elevated catecholamine levels, as well as the elevated of cortizol level, which potentiates the action of catecholamines on the vascular wall, the reduced fibrinolysis and the increased platelets aggregation – all these factors lead to a rise in the probability of rupture of the plaque and thrombosis.

Non-transmural infarctions of the myocardium do not have the same circadian characteristic. There is a characteristic peak in the late evening hours.

Circadian Characteristic of Sudden Cardiac Death

Epidemiological studies indicate that the ischaemic heart disease, acute MI accordingly, is the most frequent reason for sudden cardiac death (SCD). These finds give grounds to assume that the circadian model of SCD is similar to that in the case of MI and transient myocardial ischaemia. This assumption was confirmed for the first time in Massathusetts, 1983 [29]. The Framingham Heart Study [30] indicates 70% higher risk of SCD between 7 and 9 a.m. than in the other hours of 24-hour cycle; Population Beta Blocker Heart Attack (BHAT) [31] confirmed this circadian model. The frequency of the incidents is the highest in the morning, with second smaller peak in the evening. Gallerani, Manfredini, Ricci [32] have found the same circadian model with outpatient SCD connected with fatal arrhythmias and myocardial infarction.

Mechanisms Determining the Circadian Rhythm of the Myocardial Ischaemia

CVD have almost identical circadian characteristics. Probably the reasons for this similar circadian rhythm are the same. Most authors associate the risk of cardiovascular incidents in the morning with the rise in the physical and mental activity during these hours. Indeed, physical and psychological stress constitutes triggering factors responsible for the time of onset, the degree and the duration of the myocardial ischaemia. However, the evaluation of the degree of the physical and mental activity reveals that activity identical in degree induces myocardial ischaemia of different degree and duration depending on the hours of impact. The activity in the morning and late in the afternoon causes the most severe and prolonged ischaemia. This fact is evidence of the existence of exogenous and endogenous factors responsible for the circadian model of the myocardial ischaemia [33]. In the morning hours, a set of factors connected with prevalence of the sympathetic tone, which are a risk for cardiovascular incidents, is observed. The myocardial oxygen demand in the morning hours is the highest due to the high cardiac frequency, high arterial pressure and maximum contractility.

Atherosclerotic vessels, as well as the vessels with hypertension-induced disordered morphology of the wall, are exceptionally sensitive to vasoconstrictor stimuli. It was shown already that the plasma levels of noradrenalin and adrenaline, of renin and of cortizol are factors leading directly or indirectly to vasoconstriction, and their plasma levels have a peak in the morning hours. They raise the tone of the coronary vessels and lead to reduced supply of the myocardium with oxygen in the morning. Higher coronary vascular tone in the morning hours has been identified through angiography as well [34]; it has also been confirmed with the lower dose of vasoconstrictor drug, which is sufficient for provoking coronary spasm in the morning [35]. The factors assisting thrombi formation also have maximum activity in the morning, including the increased platelets aggregation, reduced fibrinolysis and increased blood viscosity.

The spectral components of HRV have been studied and analyzed hour by hour in patients with angina pectoris who demonstrate a sharp reduction of HF and increase of the LF/HF ratio compared to the healthy control. This result presupposes reduction of the tone of the parasympathicus in the patients with angina pectoris. The administration of isosorbide-5-mononitrate (IS-5-MN) normalizes the HF values, which presupposes restoration of the parasympathetic tone, probably due to reduction of the ischaemia. Metoprolol has the same effect on HF, but at the same time it normalizes the LF/HF ratio, which presupposes also inhibition of the existing hypersympathicotony [36]. Other authors [37] have studied the circadian characteristic in the ischaemic episodes in patients with angina pectoris or with MI. They have found high values of the LF and HF components during the night and high values for the LF/HF ratio during the day, whereby the values are the highest in the morning hours.

The ischaemic episodes in the morning and in the case of myocardial ischaemia are the most frequent. However, in the patients who had survived MI, this circadian characteristic of the ischaemia and of HRV is strongly disordered or lost. Probably the loss of the circadian characteristic of the ischaemic episodes can be explained with the disordered autonomic balance.

The circadian characteristic of the autonomic balance in patients who had survived MI, studied using HRV, probably has prognostic value as well. Malik, Farrell, Camm [38] have studied two groups of patients with MI. One of the groups is with severe post-infarction complications in the first six months, the other one is without such complications. The authors have found that low-risk patients have preserved a circadian characteristic in HRV, whereas the high-risk individuals have lost it.

It is known that in HF there is hypersympathicotony as a compensatory mechanism for improving the perfusion of the vitally important organs. This change in the vegetative balance leads to reduced HRV. However, it is reported that the circadian nature of HRV is preserved in HF [39], albeit with reduced variability. Other authors [40] report that the patients with HF are divided into two groups in terms of the circadian nature of the parameters reflecting the vagal tone: with preserved and with missing circadian nature. In the group with normal circadian characteristic of the vagal tone there are no tachyarrhythmic incidents in the morning, while in the group with missing circadian nature these incidents show a morning peak.

Circadian Characteristic of Brain Stroke

Circadian Characteristic of Brain Stroke is similar to that in the case of MI and sudden cardiac death, whereby the peak of the incidents is found to be between 6 a.m. and noon.

Loss of the circadian nature of the autonomic balance is also observed in patients with ischaemic cerebral stroke during the early period [41]. The circadian characteristic of the autonomic balance, typical of healthy individuals, with prevalence of parasympathetic activity during the night and a higher tone of the sympathicus during the day, is restored after approximately six months [41].

REFERENCES

[1] Konopka, RG; Benzer, S. Clock mutants of Drosophila melongaster. *Proc Nat Acad Sci USA*, 1971, 68, 2112-2116.

[2] Bardgiello, TA; Jackson, FR; Yong, MW. Restoration of circulation behavioural rhythms be gene transfer in drosophila. *Nature*, 1984, 312-752.

[3] Hastings M. The brain, circadian rhythms, and clock genes. *BMJ*, 1998, 317, 1704-1707.

[4] Turton, MB; Deegan T. Circadian variations of plasma cateholamine, cortisol and immunoreactive insulin concentrations in supine subjects. *Clinica Chimica Acta*, 1974, 55, 389-397.

[5] Gordon, RD; Wolfe, LK; Island, DR; Liddle. GW. A diurnal rhythm in plasma renin activity in man. *J Clin Invest*, 1966, 45, 1587-1592.

[6] Brezinski, DA; Tofler, GH; Muller, JE. Morning increase in platelet aggregability. Association with assumption of the upright posture. *Circulation*, 1988, 78, 35-40.

[7] Ehrly, AM; Jung, G. Circadian rhythm of human blood viscosity. *Biorheology*, 1973, 10, 577-583.

[8] Masuda, T; Ogawa, H; Miyao, Y. Circadian variation in fibrimolytic activity in patients with variant angina. *Br Heart Jurnal*, 1994, 71, 156-161.

[9] Massin, M; Maeyns, K, Withofs, N. Circadian rhythm of heart rate and heart rate variability. *Arch Dis Child*, 2000, 83, 179-182.

[10] Malpas, SC; Purdie, GL. Circadian variation of heart rate variability. *Cardiovasc Res*, 1990, 24, 210-213.

[11] Bonnemeier, H; Richardt, G; Potratz, J. Circadian profile of cardiac autonomic nervous modulation in healthy subjects: differing effects of aging and gender on heart rate variability. *J Cardiovasc Electrophisiol*, 2003, 14, 800-802.

[12] Yamasaki, Y; Kodama, M; Matsuhisa, M; Kishimoto, M; Ozaki, H; Tani, A; Ueda, N; Ishida, Y; Kamada, T. Diurnal heart rate variability in healthy subjects: effects of aging and sex difference. *Am J Physiol*, 1996, 271, H303-H310.

[13] Pisaruk, AV. Autonomic control of cardiovascular system in ageing: circadian rhythms. 2nd HRV non-stop congress. Available from: URL: *www.hrvcongres. org/second/*.
[14] Sten, M; Panagiotis, N; Tuck, ML. Plasma norepinephrine levels influenced by sodium intake, glucocorticoid administration, and circadian changes in normal man. *J Clin Endocrinol Metab*, 1980, 51, 1340-1345.
[15] Maling, TJB; Dollery, CT; Hamilton, CA. Clonidine and sympathetic activity during sleep. *Clin Sci*, 1979, 57, 509-514.
[16] Anderson, EA; Sinkey, CA; Lawton, WJ. Elevated sympathetic nerve activity in borderline hypertensive humans. Evidence from direct intraneural recordings. *Hypertension*, 1989, 14, 177-183.
[17] Kohara, K; Nishida, K; Maguchi, M. Autonomic nervous function in non-dipper essential hypertensives subjects. Evaluation by power spectral analysis of heart rate variability. *Hypertension*, 1995, 26, 808-814.
[18] Gusetti, S; Dassi, S; Pecis, M. Altered pattern of circadian neural control of heart period in mild hypertension. *J Hypertension*, 1991, 9, 831-838.
[19] Muiesan, ML; Rizzoni, D; Zuli, R. Circadian changes of power spectral analysis of heart rate in hypertensive patients with left ventricular hypertrophy. *High blood press*, 1996, 5, 166-172.
[20] Cakko, S; Mulingtapang, RF; Huikury, HV. Alteration in heart rate variability and its circadian rhythm in hypertensive patients with left ventricular hypertrophy free of coronary artery disease. *Am Heart J*, 1993, 126, 1364-1372.
[21] Rocco, MB; Barry, G; Campbell, S. Circadian variation of transient myocardial ishcmia in patient with coronary artery diseases. *Circulation*, 1987, 75, 395-400.
[22] Parker, JD; Testa, MA; Jimenez, AH. Morning increase in ambulatory ischemia in patients with stable coronary artery disease. Importance of physical activity and increased cardiac demand. *Circulation*, 1994, 89, 604-614.
[23] Burchinskii, GI. The role of V.P. Obraztsov and N.D. Strazhesko in the development of the ischemic heart disease problem (in Russian). *Vrach Delo*, 1989, 12, 4-6.
[24] Muller, JE; Tofler, GH; Stone, PH. Circadian variation and triggers of onset of acute cardiovascular disease. *Circulation*, 1989, 79, 733-743.
[25] Muller, JE; Stone, PH; Turi, ZG. Circadian variation in the frequency of onset of acute myocardial infarction. *N Engl J Med*, 1985, 313, 1315-1322.
[26] Willich, SN; Linderer, T; Wegscheider, K. Increased morning incidence of myocardial infarction in the ISAM study: Absence with prior beta-adrenergic blockade. *Circulation*, 1989, 80, 853-858.
[27] Ridker, PM; Mason, JE; Buring, J. Circadian variation of acute myocardial infarction and the effect of low-dose aspirin in a randomized trial of physicians. *Circulation*, 1990, 82, 897-901.
[28] Goldberg, RG; Brady, P; Muller, JE. Time of onset of symptoms of acute myocardial infarction. *Am J Cardiol*, 1990, 66, 140-144.
[29] Muller, JE; Ludmer, PL; Willich, SN. Circadian variation in frequency of sudden cardiac death. *Circulation*, 1987, 75, 131-138.
[30] Willich, SN; Levy, D; Rocco, MB. Circadian variation in the incidence of sudden cardiac death in the Framingham Heart Study populacion. *Am J Cardiol*, 1987, 60, 801-806.

[31] Peters, RW; Muller, JE; Goldstein, S. Propranolol and the morning increase in the frequency of sudden cardiac death. *Am J Cardiol*, 1989, 63, 1518-1520.

[32] Gallerani, M; Manfredini, R; Ricci, L. Sudden death may show a circadien time of risk depending on its anatomoclinical causes and age. *Jpn Heart J*, 1993, 34, 729-739.

[33] David, SK; Willem, JK; Frances, HG. Circadian variation of ambulatory myocardial ischemia. *Circulation*, 1996, 93, 1364-1371.

[34] Yasye, H; Omote, S; Takizawa, A. Circadian variation of exercise capacity in patients with Prinzmetal variant angina: role of exercise – induced coronary arterial spasm. *Circulation*, 1979, 59, 938-948.

[35] Waters, DD; Miller, DD; Bouchard, A. Circadian variation in variant angina. *Am J Cardiol*, 1984, 54, 61-64.

[36] Wennerblom, B; Lurje, L; Karlsson, T. Circadian variation of heart rate variability and rate of autonomic change in the morning hours in healthy subjects and angina patients. *Int J Cardiol*, 2001, 79, 61-69.

[37] Marchant, B; Stevenson, R; Vaishnav, S. Influence of autonomic nervous system on circadian patterns of myocardial ischaemia: comparison of stable angina with the early postinfarction period. *Br Heart J*, 1994, 71, 329-333.

[38] Malik, M; Farrell, T; Camm, J. Circadian rhythm of heart rate variability after acute myocardial infarction and its influence on the prognostic value of heart rate variability. *Am J Cardiol*, 1990, 66, 1049-1054.

[39] Adamopulos, S; Ponikovski, P; Cerquetani, E. Circadian pattern of heart rate variability in heart failure patients. Effect of physical training. *Eur Heart J*, 1995, 16, 1380-1386.

[40] Fries, R; Hein, S; Konig, J. Reversed circadian rhythms of heart rate variability and morning peak occurrence of sustained ventricular tachyarrhythmias in patient with implanted cardioverter defibrillator. *Med Sci Monit*, 2002, 8, 751-756.

[41] Korpelainen, JT; Sotaniemi, KA; Huikury, HV. Abnormal heart rate variability as a manifestation of autonomic dysfunction in hemispheric brain infarction. *Stroke*, 1997, 27, 2059-2063.

[42] Peters, RW; Muller, JE; Goldstein, S. Propranolol and the morning increase in the frequency of sudden cardiac death. *Am J Cardiol*, 1989, 63, 1518-1520.

Chapter 5

TIME-RELATED HEART AUTONOMIC BALANCE CHARACTERISTICS IN HEALTHY SUBJECTS

Mikhail Matveev
Centre of Biomedical Engineering, Bulgarian Academy of Sciences, Bulgaria
Rada Prokopova
St. Anna University Hospital, Sofia, Bulgaria

INTRODUCTION

The ANS balance changes round the clock [1, 2, 3]. In Chapter Four we pointed out that the origins of the autonomic control circadian changes are still not completely elucidated. Again there we discussed the possible hypotheses on the origin of the circadian characteristic of the heart autonomic balance (HAB).

The data on the circadian changes in HRV from the 24-hour ECG recordings are not unambiguous. Usually the LF and HF indices and their ratio LF/HF are considered as corresponding directly to the sympathetic and parasympathetic parts of ANS function. Some studies have shown circadian changes in healthy subjects with hypersympatheticotonia, from prevalence of the LF spectral range and higher LF/HF ratios in the morning [4]. Other changes have been described: low LF values during the night, increasing in the morning and remaining high during the day [5, 6]. Wennerblum, Lurje, Karlsson, et al. [7] have found low parasympathetic tone with considerable reduction of the HF component in the morning hours.

The above-cited studies confirm the existence of time-related changes in the HAB. Due to discrepancy in the tendencies of these changes derived from 24-hour ECG records, we undertook to study HAB changes in healthy subjects in two morning and afternoon hour intervals. This choice was motivated by the proven high morning risk of cardiovascular incidents and the relatively stable HAB in the afternoon hours. Studying the time-related characteristics of the autonomic balance in healthy subjects would contribute to clarifying the cardiac risk genesis. The profile of the time-related characteristics of the HAB in healthy subjects could be used for comparing the respective profiles from subjects with cardiovascular diseases, thus obtaining a better assessment of the respective disease severity.

The study (e.g. [8, 9]) included 22 healthy subjects, 11 men and 11 women, mean age 47.3 years, with no anamnesis data of cardiovascular and neurological disease, organ insufficiency and diabetes. Their previous ECG and blood pressure Holter recordings did not reveal arterial hypertension and ishaemic heart disease. In order to detect small changes in HAB, we used R-R interval recordings during resting state (RS) and in VNS stimulation with handgrip test and Valsalva manoeuvre. HRV was studied using indices obtained from ECG recordings according to the methods described in Chapter Three.

We used an adequate statistical criterion – the sign test (e.g. [10]) – to evaluate morning and afternoon differences in HAB. The sign test allows evaluation of the significance of HRV indices tendencies toward increase or decrease, by comparing two sets of results. It has the advantage over other tests (for example, the Student's t-test), of being non-parametric, i.e. free of limitations concerning the distribution of the indices values. In our specific case, this advantage is of major importance: the distributions of the HRV indices are different, some deviating considerably from the normal distribution.

The theoretical base of the variant of the test used can be presented briefly in the following way. Let N pairs of measurements v_i and v_i' ($i = 1,2...,N$) be taken for the index V in N subjects in two tests. These measurements are made for searching a possible influence of a given factor in these two tests. The values of the index for the i-th measurement with the two tests are random values V_i and V_i', and the N successive observations are on independent objects. The null hypothesis H_0 is defined as insignificance of the differences $r_i = v_i - v_i'$, i.e., the differences are due to random deviations and are not connected with the power of the index to respond to changes of the investigated factor. The alternative hypothesis H_1 is that the differences between v_i and v_i' are significant and show a different influence of the factor in the two tests.

Therefore, with H_0 it is assumed that in each measurement the random quantities V_i and V_i' have the same distribution law:

$$P(V_i < v) = P(V_i' < v) = P_i(v).$$

Consequently, the differences $V_i - V_i' = R_i$ are symmetrically distributed around zero, i.e. the deviations with (+) and (-) signs are of equal probability. The string of the number of signs N of the differences $v_i - v_i' = r_i$ consists of successive independent measurements with two possible outputs, (+) and (-), with a probability for each output

$$P[(+)] = P[(-)] = \tfrac{1}{2},$$

where the absolute differences are meaningless and the zero-differences $r_i = 0$ are excluded. If the number of signs (+) is $k_N(+)$, and the number of signs (-) is $N - k_N(+) = k_N(-)$, then testing the null hypothesis H_0 is reduced to checking the confidence of the difference between the frequency $F_N(+) = k_N(+)/N$ and the hypothetical probability $P[(+)] = \tfrac{1}{2}$.

If there are no reasons for prevailing occurrence of (+) or (-) signs, the critical domain of high absolute value deviations from $P[(+)] = \tfrac{1}{2}$ is taken for confidence level q/100. For the purpose, a number δ is selected, such as to satisfy the relation

$P[|k_N(+) - N/2| >= \delta] = P[k_N(+) >= N/2 + \delta] + P[k_N(+) <= N/2 + \delta] = p1 + p2 = q/100$.

Obviously, if m_N (the lesser of $k_N(+)$ and $k_N(-)$) satisfies the equality $m_N \leq N/2-\delta$, one of the inequalities $k_N(+) \leq N/2-\delta$ (if $k_N(+) = m_N$ is smaller) or $k_N(+) \leq N/2-\delta$ (if $k_N(-) = m_N$ is smaller) is satisfied. Then, according to the rule of addition of independent events, for $m_N = \min[k_N(+), k_N(-)]$, we obtain

$P(m_N <= N/2 - \delta) = p1 + p2 = q/100$.

The critical integer numbers **m_N**, for which the null hypothesis H_0 is accepted (if $m_N = \min[k_N(+), k_N(-)] \leq$ **m_N**), or rejected (if $m_N = \min[k_N(+), k_N(-)] >$ **m_N**), at a confidence level $p = q/100$ (e.g. $p = 0.05$) and a number of measurements N, are taken from a table (e.g. [10]).

COMPARISON BETWEEN MORNING AND AFTERNOON VALUES OF HRV INDICES DURING RESTING STATE, HANDGRIP TEST AND VALSALVA MANOEUVRE

The results obtained with application of the sign test for checking the significance of the differences in the HRV indices, measured in the morning and in the afternoon, during resting state and with handgrip test and Valsalva manoeuvre, are shown in Tables 5.1, 5.2 and 5.3. In these tables, columns $\kappa_N(+)$, $\kappa_N(-)$ and **m_N** contain accordingly the number of deviations with sign (+), sign (-) and the critical number of which should not be exceeded by the smaller one of $\kappa_N(+)$ and $\kappa_N(-)$ so that there would be a significant ($p<0.05$) change in the values of a concrete parameter in the morning and in the afternoon measurements for the number of persons with non-zero difference during the two measurements indicated in column N. The signs are determined depending on whether the difference between the value in the morning minus the value in the afternoon is positive or negative. The third column in the table indicates the confidence level p, which corresponds to the smaller value among $\kappa_N(+)$ and $\kappa_N(-)$ and the number of persons N.

The results show no significant changes between morning and afternoon HRV indices measured with all three tests. The distances of each of the indices from the level of significant difference are shown in Figures 5.1, 5.2 and 5.3, by comparing the smaller of $k_N(+)$ and $k_N(-)$ with **m_N**. The area of the significant differences is circled in the figures.

Table 5.1. Sign test – HRV indices during resting state: morning and afternoon measurements.

Indices	N	$p =$	$k_N(-)$	$k_N(+)$	Critical numbers m_N
SDRR	21	1.0000	10	11	5
MDRR	22	0.8312	9	13	5
dSDRR	22	0.5224	13	9	5
dMDRR	22	0.5224	13	9	5
PNN50	19	0.3588	11	8	4
RMSSD	22	0.5224	13	9	5
VLF	22	0.1356	6	16	5
LF	22	0.5224	8	14	5
HF	22	0.8312	12	10	5
LF/HF	22	0.5224	8	14	5

Table 5.2. Sign test – HRV indices with handgrip test: morning and afternoon measurements.

Indices	N	$p =$	$k_N(-)$	$k_N(+)$	Critical numbers m_N
SDRR	22	0.8312	9	13	5
MDRR	21	0.6625	8	13	5
dSDRR	20	0.5023	8	12	5
dMDRR	20	0.8231	9	11	5
PNN50	17	1.0000	7	10	4
RMSSD	21	0.6625	8	13	5
VLF	22	0.8312	11	11	5
LF	22	0.8312	11	11	5
HF	22	0.1356	6	16	5
LF/HF	22	0.0550	15	7	5

Table 5.3. Sign test – HRV indices with Valsalva manoeuvre: morning and afternoon measurements.

Indices	N	$p =$	$k_N(-)$	$k_N(+)$	Critical numbers m_N
SDRR	20	0.8230	9	11	5
MDRR	22	0.2864	8	14	5
dSDRR	22	0.8311	9	13	5
dMDRR	19	0.3587	7	12	4
PNN50	19	0.3587	7	12	4
RMSSD	22	0.8311	9	13	5
LF	22	0.8311	9	13	5
HF	22	0.5224	8	14	5
LF/HF	22	0.8311	11	11	5

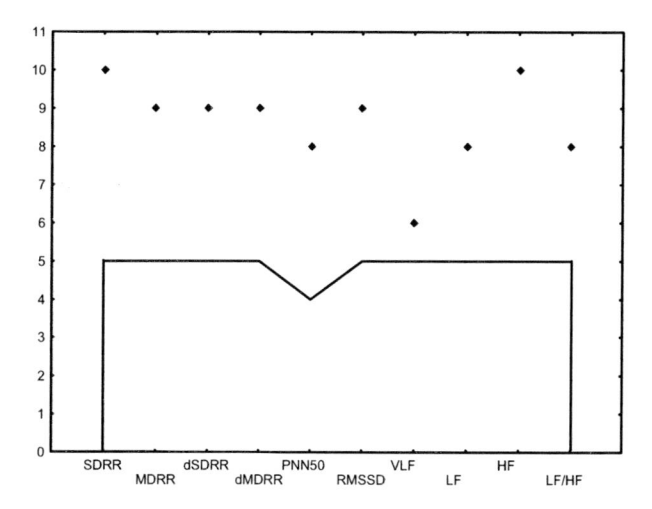

Figure 5.1. Sign test: HRV indices in the resting state; comparison between morning and afternoon measurements.

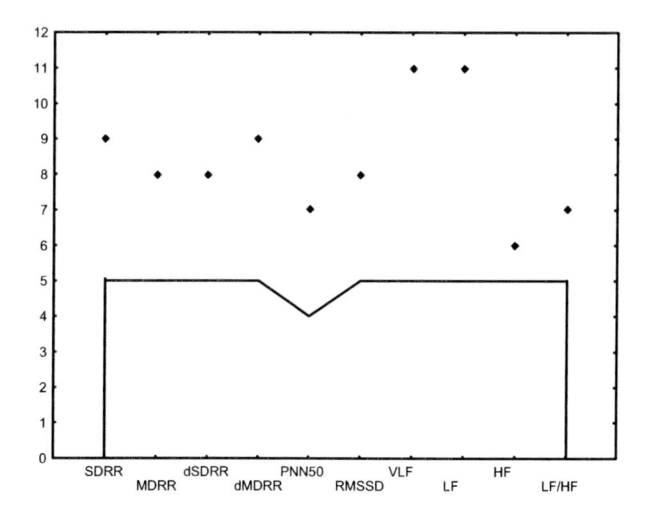

Figure 5.2. Sign test: HRV indices in the handgrip test; comparison between morning and afternoon measurements.

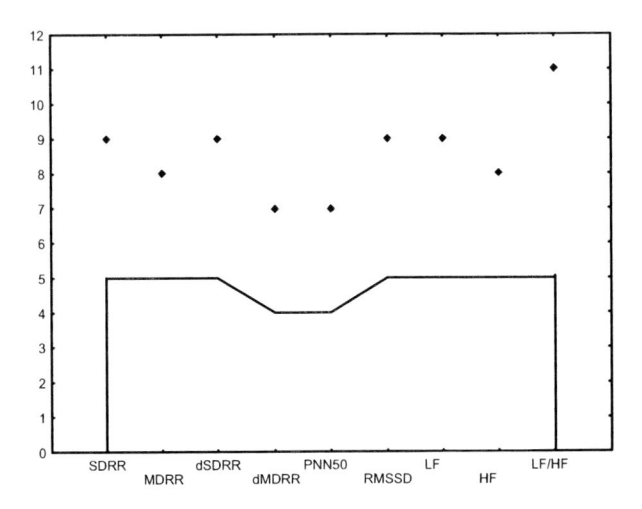

Figure 5.3. Sign test: HRV indices in the Valsalva manoeuvre; comparison between morning and afternoon measurements.

COMPARISON BETWEEN VALUES OF HRV INDICES DURING RESTING STATE AND STIMULATION TESTS

Morning Measurements

Table 5.4 and Figure 5.4 contain the results of checking the significance of the differences between HRV indices values in comparing morning measurements during resting state and during the handgrip test. Table 5.5 and Figure 5.5 contain the results of the sign test for checking the significance of the differences between HRV indices values in comparing morning measurements during resting state and Valsalva manoeuvre.

Comparing the resting state with the handgrip test, significant changes ($p<0.05$) can be seen in two time-domain indices (SDRR and MDRR) and in the three frequency-domain indices (VLF, LF and HF). The handgrip test values of LF and HF decreased with respect to the resting state values, with a highly significant difference ($p<0.0001$). The changes in VLF were similar: for 21 subjects out of 22 there are lower values in the handgrip test compared to resting state.

There were different results in the comparison of the resting state with the Valsalva manoeuvre. There is a significant change in all time-domain indices. The frequency-domain indices LF and HF are again of lesser values compared to resting state. For HF, changes were found in all subjects. (The Valsalva manoeuvre did not include VLF, due to the short ECG record duration). The LF/HF index did not belong to the constellations, for the two types of comparison: between resting state and handgrip test and between resting state and Valsalva manoeuvre.

Table 5.4. Sign test – HRV indices during resting state and with the handgrip test: morning measurements (significantly different indices ($p<0.05$) are marked by *).

Indices	N	$p =$	$k_N(-)$	$k_N(+)$	Critical numbers $\underline{m}_N$
SDRR*	20	0.0139	4	16	5
MDRR*	21	0.0291	5	16	5
dSDRR	19	0.1687	6	13	4
dMDRR	19	0.1687	6	13	4
PNN50	16	0.2113	5	11	3
RMSSD	22	0.2864	8	14	5
VLF*	22	0.0001	1	21	5
LF*	22	0.0000	0	22	5
HF*	22	0.0000	0	22	5
LF/HF	22	0.2864	8	14	5

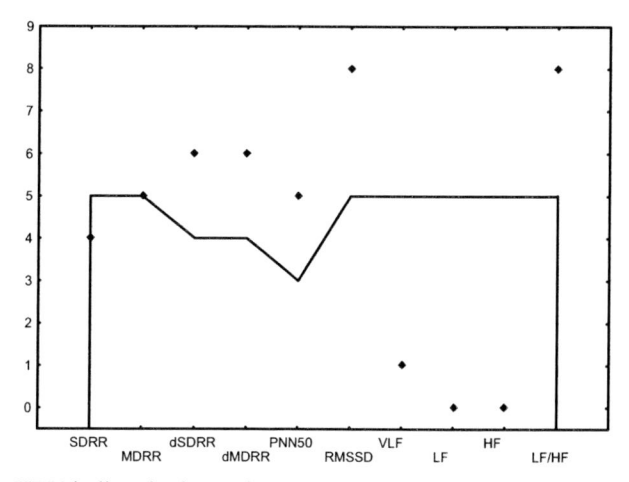

Figure 5.4. Sign test: HRV indices in the resting state and with the handgrip test: comparison between morning measurements.

Table 5.5. Sign test – HRV indices during resting state and with the Valsalva manoeuvre: morning measurements (significantly different indices ($p<0.05$) are marked by *).

Indices	N	$p =$	$k_N(-)$	$k_N(+)$	Critical numbers $\underline{m}_N$
SDRR*	22	0.0014	19	3	5
MDRR*	22	0.0014	19	3	5
dSDRR*	22	0.0003	20	2	5
dMDRR*	22	0.0056	18	4	5
PNN50*	21	0.0088	17	4	5
RMSSD*	22	0.0056	18	4	5
LF*	22	0.0014	3	19	5
HF*	22	0.0000	0	22	5
LF/HF	22	0.8312	11	11	5

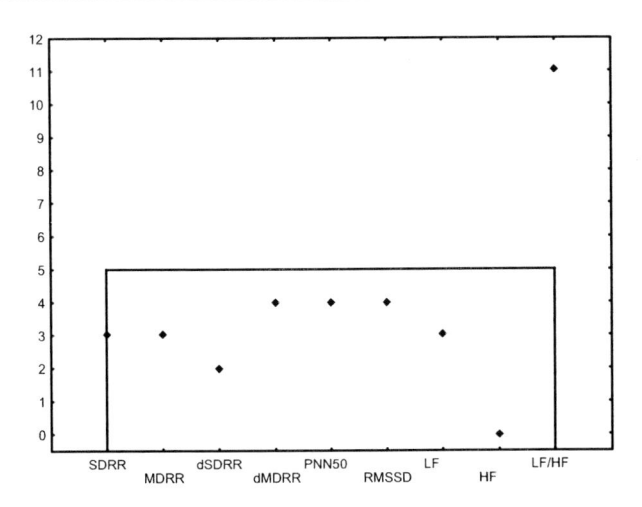

Figure 5.5. Sign test: HRV indices in the resting state and with the Valsalva manoeuvre: comparison between morning measurements.

Afternoon Measurements

The time-domain indices dSDRR, dMDRR, PNN50 and RMSSD showed significant differences of the handgrip test values compared to resting state (Table 5.6 and Figure 5.6). It is interesting to note that these parameters are from the dRR-tachogram. Three of the frequency-domain indices (VLF, LF and HF) also registered significant differences. HF showed changes of the same sign for all 22 subjects studied. The LF changes were similar for 21 of the 22 subjects.

Table 5.6. Sign test – HRV indices during resting state and with the handgrip test: afternoon measurements (significantly different indices ($p<0.05$) are marked by ∗).

Indices	N	$p =$	$k_N(-)$	$k_N(+)$	Critical numbers $\underline{m}_N$
SDRR	22	0.0550	6	16	5
MDRR	20	0.1175	6	14	5
dSDRR∗	21	0.0291	5	16	5
dMDRR∗	22	0.0190	5	17	5
PNN50∗	19	0.0059	3	16	4
RMSSD∗	21	0.0088	4	17	5
VLF∗	22	0.0014	3	19	5
LF∗	22	0.0001	1	21	5
HF∗	22	0.0000	0	22	5
LF/HF	22	0.5224	13	9	5

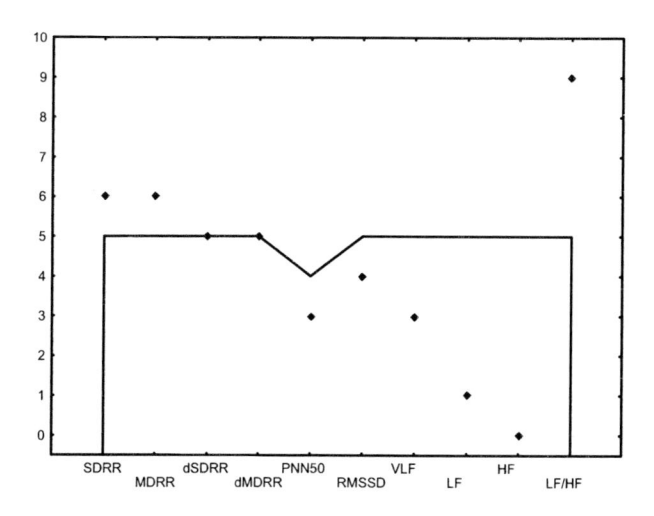

Figure 5.6. Sign test: HRV indices in the resting state and with the handgrip test: comparison between afternoon measurements.

Comparing the values during resting state with those from the Valsalva manoeuvre (Table 5.7 and Figure 5.7), only two time-domain indices showed significant difference at marginal level, namely dSDRR and dMDRR. The frequency-domain indices LF and HF showed significant changes, but of lower level compared to the handgrip test.

Similar to the morning tests, LF/HF did not show significant difference between resting state and handgrip test and between resting state and the Valsalva manoeuvre.

Table 5.7. Sign test – RRV indices during resting state and with the Valsalva manoeuvre: afternoon measurements (significantly different indices ($p<0.05$) are marked by *).

Indices	N	$p=$	$k_N(-)$	$k_N(+)$	Critical numbers $\underline{m}_N$
SDRR	22	0.1356	15	7	5
MDRR	22	0.1356	15	7	5
dSDRR*	21	0.0291	16	5	5
dMDRR*	20	0.0442	15	5	5
PNN50	20	0.2636	13	7	5
RMSSD	22	0.5224	13	9	5
LF*	22	0.0056	4	18	5
HF*	22	0.0003	2	20	5
LF/HF	22	0.2864	14	8	5

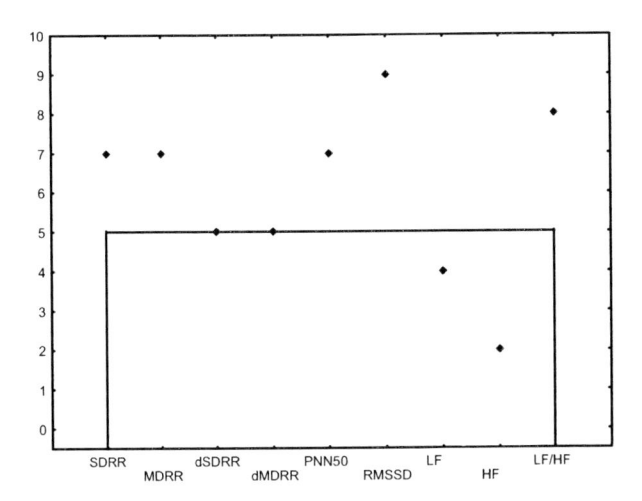

Figure 5.7. Sign test: HRV indices in the resting state and with the Valsalva manoeuvre; comparison between afternoon measurements

TIME-RELATED AUTONOMIC BALANCE CHANGES INDICATOR (TRABI)

The results of the two previous sections showed: (i) no significant differences between morning and afternoon values of HRV indices for all three types of measurements; (ii) statistically significant differences between resting state and stimulation values, but in different indices constellations for the morning and afternoon measurements. Therefore, it was justified to look for relative changes in time-related HAB indices. First, the indices values during resting state and with stimulation in morning measurements were compared. Next, the same comparison was done for afternoon measurements. The power of the HRV indices to detect the time-related differences between these two comparisons was evaluated by the *Time-Related Autonomic Balance changes Indicator (TRABI)* introduced by us [11, 12].

Let two measurements be made at two different moments of time in order to detect the influence of a factor in two tests. In each measurement i ($i = 1,2$) we obtain values of an index V for N objects in two tests. For each measurement, comparing the N parallel pairs of values of the index V, we have $k_i(+)$ positive, $k_i(-)$ negative and $k_i(0)$ zero differences. Then the following limitation is valid:

$$k_i(+) + k_i(-) + k_i(0) = N \ (i = 1,2). \tag{1}$$

Let us consider the indicator

$$\mathbf{I} = |[k_1(+) - k_1(-)] - [k_2(+) - k_2(-)]|/2N = |dk_1 - dk_2|/2N, \tag{2}$$

where, taking into account (1), the zero differences $k_i(0)$ are indirectly included through N. Obviously, the indicator introduced by (2) measures the power of the index to respond to changes in comparing results of two tests in two measurements, performed at different moments of time (for example, morning and afternoon).

Equations (1) and (2) define a range of values of [0;1] for I. In order to illustrate the conditions for obtaining the values of **I**, we assume for the sake of convenience (without influencing the essence of the results) that there are no zero differences in the two measurements. Hence,

$$k_i(0) = 0 \text{ and } N = k_i(+) + k_i(-). \tag{3}$$

(The presence of zero differences reduces the values of the indicator (2), depending on their number.)

Obviously, the indicator is zero if $dk_1 = dk_2$. Considering (3), this is possible only if both equalities $k_1(+) = k_2(+)$ and $k_1(-) = k_2(-)$ are satisfied. In other words, the indicator value is zero (no time-related changes) only in case there is *no change* in the number of positive and negative differences of the index V values obtained from two tests in the two measurements.

The maximum value **I** = 1 is defined from $|dk_1 - dk_2| = 2N$ and considering (3), it is necessary that $dk_1 = -dk_2$, or that one of the two conditions be satisfied: $\{k_1(+) = N, k_1(-) = 0; k_2(+) = 0, k_2(-) = N\}$, or $\{k_1(+) = 0, k_1(-) = N; k_2(-) = N, k_2(+) = 0\}$. There is a case of *full change* in the number of positive and negative differences of the index V values from the two measurements and the two tests.

In this study, it was convenient to divide the values of the indicator into several ranges. In order to introduce such ranges, we can use another property of **I**. Let $\alpha.100$ be a percentage of deviation of $k_i(+)$ from half of the data (N/2) and let $k_1(+)$ exceed N/2, and $k_2(+)$ does not attain N/2 with this percentage. Then $k_1(+) = N/2(1+\alpha)$, $k_2(+) = N/2(1-\alpha)$, respectively $k_1(-) = N/2(1-\alpha)$, $k_2(-) = N/2(1+\alpha)$ and **I** = α. The selected mirror symmetry for $k_i(+)$ and $k_i(-)$ allows to compose a table with values of *I* from 0 to 1, incremented by a defined step. *It should be noted that the real values of I do not exactly correspond to α, due to the fact that $k_i(\pm)$ are integer numbers and the expressions N/2(1±α) should be rounded to the nearest integer. For a small increment steps of α, the rounding may result in equal **I** values for adjacent values of α.* This specificity ought to be borne in mind when discussing the values of the index for the HRV parameters in this and the next chapters.

Table 5.8 is designed for defining TRABI values in healthy subjects, for 0.05 α-value increments. The values are given for $k_i(\pm)$, rounded to the nearest integer. For the purpose of this study, we defined four ranges of **I** values: (i) **I** = 0 [$k_i(\pm) = k_2(\pm) = N/2$] to **I** = 0.20, corresponding to zero (0) or insignificant difference between morning and afternoon results; (ii) I>0.20 to I<0.5; (iii) **I** = 0.5 to I<0.75 and (iv) I>0.75, corresponding to moderate, considerable and very high differences, respectively. It should be noted that for **I** = 0.5, the lesser of $k_i(+)$ and $k_i(-)$ is equal to 5, which for N = 22 is the critical **m_N**. For **m_N** = 5 and p = 0.05, the sign test shows significant difference between the values of a given index in the two tests (for example, resting state versus a stimulation test).

The evaluation results for the morning and afternoon measurements during resting state and with the handgrip test are shown in Table 5.9 and Figure 5.8. The respective results for resting state and Valsalva manoeuvre are shown in Table 5.10 and Figure 5.9. These two tables include the number of negative $k_i(-)$ and positive $k_i(+)$ differences during resting state and stimulation tests in the morning (i = 1) and in the afternoon (i = 2). In addition, the number of zero differences $k_i(0)$ is shown, as the latter are indirectly included in the computation of the indicator value.

The lack of significant changes in the morning and in the afternoon HRV indices, during resting state and with the two stimulation tests, can be explained with the relative stability of HAB in healthy subjects.

Changes in HAB were found after comparing resting state HRV with HRV provoked by stimulation tests. The comparison of morning results during resting state versus the handgrip test and during resting state versus the Valsalva manoeuvre showed different sensitivity of the time-domain and frequency-domain indices. We explain the small number of indices reacting to the handgrip test with respect to resting state with the morning background parasympatheticotonia which suppresses the effect of additional sympathetic stimulation. A considerably larger set of significantly differing indices in the Valsalva manoeuvre with respect to resting state can be explained with low morning parasympathetic tone, thus allowing stimulation. Penttila [13] has supported such a hypothesis. He has found considerable reduction in HF, RMSSD and PNN50 values in healthy subjects, with full vagus medicamentous blockade. The morning hypersympatheticotonia that we observed in healthy subjects might be related to higher risk of cardiovascular incidents. Aubert, Beckers and Verheyden [14] explicitly stress on the relation of reduced HRV with increased risk in clinically healthy subjects.

Table 5.8. Values of TRABI for selected mirror symmetry of $k_i(+)$ and $k_i(-)$ and for 0.05 $\tilde{\alpha}$-value increments.

N/2	α	$k_1(+)$	$k_1(-)$	$k_2(+)$	$k_2(-)$	*TRABI*
11	0.00	11	11	11	11	0.000
11	0.05	10	12	12	10	0.091
11	0.10	10	12	12	10	0.091
11	0.15	9	13	13	9	0.182
11	0.20	9	13	13	9	0.182
11	0.25	8	14	14	8	0.273
11	0.30	8	14	14	8	0.273
11	0.35	7	15	15	7	0.364
11	0.40	7	15	15	7	0.364
11	0.45	6	16	16	6	0.455
11	0.50	5	17	17	5	0.545
11	0.55	5	17	17	5	0.545
11	0.60	4	18	18	4	0.636
11	0.65	4	18	18	4	0.636
11	0.70	3	19	19	3	0.727
11	0.75	3	19	19	3	0.727
11	0.80	2	20	20	2	0.818
11	0.85	2	20	20	2	0.818
11	0.90	1	21	21	1	0.909
11	0.95	1	21	21	1	0.909
11	1.00	0	22	22	0	1.000

Table 5.9. Values of TRABI for the HRV indices in comparison of morning and afternoon measurements during resting state and with the handgrip test.

Indices	Morning measurements				Afternoon measurements				TRABI
	N	$k_1(-)$	$k_1(+)$	$k_1(0)$	N	$k_2(-)$	$k_2(+)$	$k_2(0)$	
SDRR	20	4	16	2	22	6	16	0	0.045
MDRR	21	5	16	1	20	6	14	2	0.068
dSDRR	19	6	13	3	21	5	16	1	0.091
dMDRR	19	6	13	3	22	5	17	0	0.114
PNN50	16	5	11	6	19	3	16	3	0.159
RMSSD	22	8	14	0	21	4	17	1	0.159
VLF	22	1	21	0	22	3	19	0	0.091
LF	22	0	22	0	22	1	21	0	0.045
HF	22	0	22	0	22	0	22	0	0.000
LF/HF	22	8	14	0	22	13	9	0	0.227

Table 5.10. Values of TRABI for the RRV indices in comparison of morning and afternoon measurements during resting state and with the Valsalva manoeuvre.

Indices	Morning measurements				Afternoon measurements				TRABI
	N	$k_1(-)$	$k_1(+)$	$k_1(0)$	N	$k_2(-)$	$k_2(+)$	$k_2(0)$	
SDRR	22	19	3	0	22	15	7	0	0.182
MDRR	22	19	3	0	22	15	7	0	0.182
dSDRR	22	20	2	0	21	16	5	1	0.159
dMDRR	22	18	4	0	20	15	5	2	0.091
PNN50	21	17	4	1	20	13	7	2	0.159
RMSSD	22	18	4	0	22	13	9	0	0.227
LF	22	3	19	0	22	4	18	0	0.045
HF	22	0	22	0	22	2	20	0	0.091
LF/HF	22	11	11	0	22	14	8	0	0.136

The relatively low afternoon sympathetic tone results in the opposite tendency of the time-domain indices. All six indices showed significant differences during resting state versus the handgrip test, which we relate to possible additional sympathetic stimulation. The fact they are parameters derived from the dRR-tachogram is due to their higher sensitivity to HRV, reduced by the sympathetic stimulation. The afternoon relative hypersympatheticotonia limits the possibility to provoke considerable change in HAB by the Valsalva manoeuvre. Hence, the number of significantly differing time-domain indices was reduced.

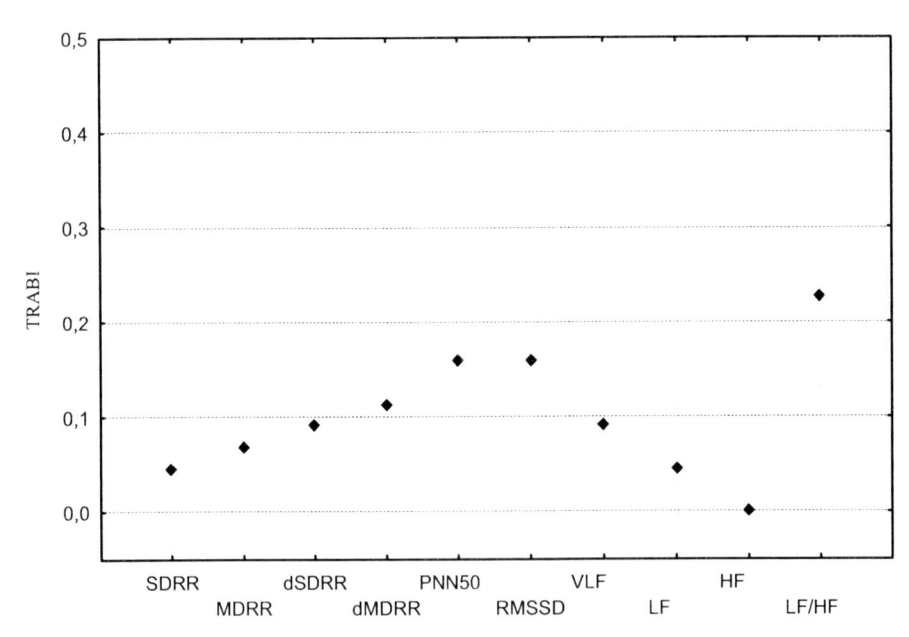

Figure 5.8. Values of TRABI for the HRV indices in a comparison of morning and afternoon measurements in the resting state and with the handgrip test.

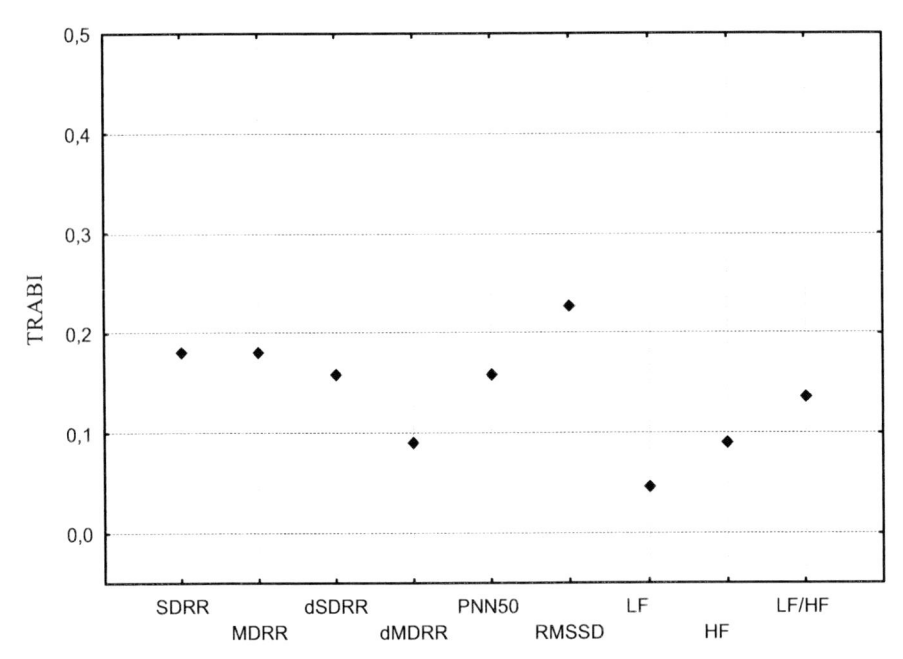

Figure 5.9. Values of TRABI for the HRV indices in a comparison of morning and afternoon measurements in the resting state and with the Valsalva manoeuvre.

The results of morning and afternoon measurements reveal the possibilities of HRV indices to respond to changes in HAB. The time-domain and frequency-domain indices have different sensitivity to the changes. On the other hand, the two sets of time-domain indices (from RR and dRR tachograms respectively) have different power to represent HAB changes.

The frequency-domain indices (VLF, LF and HF in the handgrip tests and LF and HF in the Valsalva manoeuvre) show high sensitivity to HAB changes, even considering its relative stability in healthy subjects. This is demonstrated by the unilateral changes of these indices for all or most subjects in morning and afternoon measurements. Houle and Billman [15] pay attention to the high sensitivity of the frequency-domain indices in vagal or sympathetic stimulation. Therefore, the specific *time-related balance changes are masked* in the prevailing tendency to changes in the frequency-domain indices in stimulation compared to resting state. In other words, *the frequency-domain indices are insensitive to time-related changes in the VNS*. Only LF/HF shows a different response to morning versus afternoon changes, due to its inherent elasticity as the ratio of the values of two indices. Its power for assessment of the autonomic balance during resting state and in stimulation is limited by its high variability [16].

The HRV indices power to respond to time-related changes of HAB is quantitatively represented by the values of the proposed indicator (TRABI) (Figure 5.8 and Figure 5.9).

The TRABI values corresponding to morning and afternoon differences in handgrip stimulation versus resting state divide the HRV indices into two groups with respect to their power to reflect the changes: (i) low and insignificant (SDRR, MDRR, dSDRR, dMDRR, PNN50, RMSSD, VLF and LF), and (ii) moderate (LF/HF). For the indices of the first group, the result is due to the equal tendencies of their value changes. In these cases, lower values prevail in the morning and afternoon measurements. The relations of positive and negative changes of these indices between morning and afternoon values can be explained by the already shown relative stability of the autonomic balance in healthy subjects. A certain difference can be observed for PNN50 and RMSSD: in the afternoon measurements the number of positive resting state-to-handgrip test differences increases. We explain this finding with the power of dRR-tachogram indices to respond to the relatively lower sympathetic level in the afternoon – they react to increased adjacent RR interval differences. Molgaard, Sorensen and Bjerregaard [2] also relate PNN50 changes with parasympathetic activity and imply this to be valid for other indices sensitive to parasympathetic activity. The assessments for HF and LF by TRABI confirm the assumption about the limited power of these indices to respond to time-related autonomic balance changes.

For LF/HF, the tendency is changing: the ratios of positive and negative differences in the morning and in the afternoon are opposite. For the handgrip test, the morning value is lower in 14 subjects and higher in 8, while the afternoon values are lower in 9 subjects and higher in 13. The transition to prevailingly higher values (handgrip test versus resting state) in the afternoon is due to higher LF values in stimulation. It was already commented that these higher values are possible because of the lower afternoon sympathetic tone, yielding additional stimulation.

The HRV indices are also divided into two groups by the autonomic balance indicator, with respect to their power to respond to changes in the morning and in the afternoon autonomic balance, due to the Valsalva manoeuvre stimulation, with respect to resting state balance. SDRR, MDRR, dSDRR, dMDRR, PNN50, LF, HF and LF/HF have values below 0.20 (insignificant power). The low indicator values are due to the prevailingly higher values of the indices in stimulation with respect to resting state, both for morning and afternoon measurements. In contrast to the data on the handgrip test, no close relationship can be found between positive and negative differences in morning and afternoon results. There is a tendency toward reduced number of negative differences in the afternoon. This is especially pronounced for SDRR, MDRR, and to a lesser extent for dSDRR and PNN50 (Table 5.9). This tendency might be due to the limited power of the Valsalva manoeuvre to stimulate the relatively more excited parasympathetic system in the afternoon. Molgaard, Sorensen and Bjerregaard [2] supported such an interpretation, reporting that SDRR was reduced with a suppressed sympathetic tone (i.e., with prevailing parasympathetic tone). The low indicator values for LF and HF in this case, as with the handgrip test, is due to the virtually equal ratio of morning and afternoon positive and negative differences.

The group of moderate power to represent the changes RMSSD. The reason for this parameter is a more pronounced change in the ratio of morning and afternoon positive and negative differences, but with still prevailing negative differences.

The heart autonomic balance in healthy subjects is characterized by relative stability. In spite of this relative stability, HRV indices can reveal time-related autonomic balance changes during the different time intervals of the 24-hour period.

The power of HRV indices to respond to circadian changes in the autonomic balance due to stimulation is different. The frequency-domain indices have low sensitivity for changes and do not take part in the composition of a time-related heart autonomic balance profile in healthy subjects. The time-domain indices have considerably higher power to react to circadian changes in the balance in stimulation of the vegetative nervous system components.

The proposed indicator for assessment of time-related changes is an adequate quantitative measure for the individual power of the HRV indices to respond to circadian changes in the heart autonomic balance. Specific time-related changes in subjects with cardiovascular disease in initial or advanced stage can be defined by comparison with the corresponding profile of circadian changes in healthy subjects. There is a method for identifying changes in the autonomic balance that are specific for the cardiovascular diseases.

REFERENCES

[1] Arshed, A. Circadian rythms in cardiovascular disease. *Am.Heart J.*, 1990, 120, 726-733.

[2] Molgaard, H; Sorensen, KE; Bjerregaard, P. Circadian variation and influence of risk factors on heart rate variability in healthy subjects. *Am J Cardiol*, 1991, 68, 777-84.

[3] Van Leeuwen, P; Bettermann, H; An der Heiden, U; Kumell, HC. Circadian aspects of apparent correlation dimension in human rate dynamics. *Am J Physiol*, 1995, 269, H130-134.

[4] Muiesan, ML; Rizzoni, D; Zulli, R; Castellano, M; Bettoni, G. Circadian changes of power spectral analysis of heart rate in hypertensives patients with left ventricular hyperthropfy. *High Blood Press*, 1996, 5, 166-172.

[5] Chakko, S; Mulingtapang, RF; Huikuri, HV; Kessler, KM; Materson, BJ; Myerburg, RJ. Alteration in heart rate variabilty and its circadian rhythm in hypertensife patients with left ventricular hyperthrophy free of coronary artery disease. *Am Heart J*, 1993, 126, 1364-1372.

[6] Guzzetti, S; Dassi, S; Pecis, M. Altered pattern of circadian neural control of heart period in mild hypertension. *J Hypertens*, 1991, 9, 831-838.

[7] Wennerblum, B; Lurje, L; Karlsson, T; Tygesen, H; Vahisalo, R; Hjalmarson, A. Circadian Variations of Heart Rate Variabilty and the role of autonomic change in morning hours in healthy subjects and angina patients. *Int J Cardiol*, 2001, 79(1), 61-69.

[8] Matveev, M; Prokopova, R; Nachev, Ch. Time-related heart autonomic balance characteristics in healthy subjects. *Physiol. Meas.*, 2003, 24, 727-743.

[9] Prokopova, R., Matveev, M. Time-related vegetative balance characteristics. *Proceedings of the 19th European congress of the International Society of Non-Invasive Cardiology*. Istanbul, Turkey, May, 2003, PP-30.

[10] Conover, WJ. *Practical Nonparametric Statistics*. 2nd ed. NY: John Wiley & Sons; 1980.

[11] Matveev, M; Prokopova, R; Nachev, Ch. Time-Related Vegetative Balance Characteristics in Healthy Subjects. *Computers in Cardiology*, 2003, 30, 457-460.

[12] Matveev, M. Non-parametric sign criterion for estimation of the sensitivity of object's features to influences of a factor. *Proceedings of the Ninth national conference with international participation on biomedical physics and engineering*. Sofia, Bulgaria, October, 2004, 180-185.

[13] Penttila, J. Heart Rate Variability and Baroreflex Sensitivity in the Assessment of Anticholinergic Drug Effect – Studies on Healthy Volunteers. *Ann Univ Turk*, 2002, D495.

[14] Aubert, AE; Beckers, F; Verheyden, B. Heart rate Variability. Methods and Applications. *Proceedings of the 19th European congress of the International Society of Non-Invasive Cardiology*. Istanbul, Turkey, May, 2003, 144-146.

[15] Houle, MS; Billman, GE. Low-frequency component of the heart rate variability spectrum: a poor marker of sympathetic activity. *Am J Physiol*, 1999, 276, H215-223.

[16] Hojgaard, MV; Holstein-Rathlou, NH; Agner, E; Kanters, JK. Dynamics of spectral components of heart rate variability during changes in autonomic balance. *Am J Physiol*, 1998, 275, H213-219.

Chapter 6

CHANGES IN THE AUTONOMIC CARDIAC CONTROL IN ARTERIAL HYPERTENSION

Rada Prokopova
St. Anna University Hospital, Sofia, Bulgaria
Mikhail Matveev
Centre of Biomedical Engineering, Bulgarian Academy of Sciences, Bulgaria

AUTONOMIC BALANCE IN HYPERTENSION

Arterial hypertension is the most significant risk factor for the development of vascular disorders in the target organs: heart, brain, kidneys and limbs. It is connected both with the occurrence and with the progression of the cardiovascular diseases. In large-scale studies, e.g., the Framingham Heart Study [1], it is emphasized that after the age of 55 years, 90% of the normotensive individuals become hypertensive patients.

The severity of the organ damage and of the cardiovascular incidents depends on the degree of the hypertensive disease. For this reason, the 7th JNC report [2] introduces the concept of pre-hypertension for arterial pressure between 120/80 and 140/90. Recommendations are given on the applying of therapy for some concomitant changes (e.g., existence of other risk factors). The high incidence of the hypertensive disease and its severe consequences for the entire organism are the reason for the heightened interest in the principal pathophysiological disorders observed in mild forms of the hypertensive disease. Their elucidation would provide an opportunity for screening of the mild hypertensive patients and for determining an adequate therapeutic intervention already at that initial stage of arterial hypertension.

The participation of the sympathetic nervous system in the onset and progression of the hypertensive disease has been studied by many researchers. The hyperactivity of the sympathetic nervous system stimulates RAAS, intensifies the sodium absorption, increases the heart rate, the stroke volume and the peripheral vascular resistance, and through this mechanism it provokes a rise in the arterial pressure [3]. The hypersympatheticotony in hypertensive patients leads to other severe consequences as well. There exists evidence that it disturbs the glucose delivery to the muscles, delays the postprandial lipids clearance in the liver and leads to hyperlipidaemia. The activation of the sympathicus in the heart is associated with a higher risk of SCD. The trophic effect of the sympathetic activation on the cardiac muscle, leading to left-ventricular hypertrophy, is also known [4]. The sympathetic hyperactivity is also connected with vascular remodeling and atherosclerosis [5].

Anderson, Sinkey, Lawton, et al. [6] have proven the existence of elevated activity of the sympathicus in the case of borderline hypertension through measurement of the muscle nervous activity. Through the use of the frequency indices of HRV, Guzzetti, Piccaluga, Casati, et al. [7] also find sympathetic hyperactivity in the autonomic balance.

As the hypersympatheticotony is connected with the progression of the left-ventricular hypertrophy, it would be natural for many of the studies to be aimed at evaluating the autonomic balance in hypertensive patients with and without left-ventricular hypertrophy (LVH). Petretta, Bianchi, Marchiano, et al. [8] have studied the influence of the autonomic balance through the frequency indices of HRV in hypertensive patients with and without LVH and in normotensive individuals. They have discovered higher values for LF and HF in the normotensive individuals compared to the values in the hypertensive patients, as well as progressively diminishing values for these indices from hypertensive patients without LVH to hypertensive patients with LVH. In [9] the authors have proven that the left-ventricular mass index is connected with the body mass index, with VLF, with LF and with the systolic arterial pressure in hypertensive patients.

In some studies it is claimed that the patients with mild hypertension lack disorders in the autonomic balance of the heart and that the balance is disturbed only in the case of severe hypertension. Mussalo, Vanninen, Ikaheimo, et al. [10] compare the autonomic balance in patients with mild and severe essential hypertension and normotensive individuals. They find that only patients with severe arterial hypertension demonstrate reduced values of the indices from the time- and frequency-domain. Other authors [11] study HRV in patients with borderline hypertension without and with LH, and compare it with the variability in healthy subjects. On the basis of data from 24 ECG recordings, they have found that the patients with LVH have reduced values for the indices from the time domain SDNN and SDANN. At the same time, there is no change in the values of these indices in the patients without LVH compared to the values in healthy subjects. However, these results differ from the results obtained in large-scale studies – see, e.g., [1]. In that study, HRV is used as a marker for the state of the autonomic function of the heart in normotensive individuals and in hypertensive patients. The possibility is also checked for HRV to be a predictor for the appearance of hypertension. The authors have found that all indices for HRV are significantly lower for the hypertensive individuals. They have also proven that in normotensive men the low HRV is associated with higher risk of development of hypertension. The low values of LF improves the prediction of risk of hypertension. *These findings presuppose that disturbed autonomic regulation is discovered already in the early stages of the hypertensive disease.*

It is known that higher than normal blood pressure is connected with higher cardiovascular risk. The identification of individuals with higher risk of cardiovascular incidents will improve the long-term prognosis concerning them. Lucini, Mela, Malliani, et al. [12] study whether the subjects with high normal arterial pressure already have damaged autonomic control of the heart and vessels. The autonomic regulation is studied through spectral analysis of HRV, the variability of the systolic arterial pressure, the changes in the variability and pressure during rest and after standing up. The authors have found that the patients with high normal arterial pressure have significant disorders in the sinoatrial regulation. The persons with hypertension demonstrate disordered arterial regulation as well. With sympathetic stimulation (standing up), the indices for HRV are progressively reduced, whereby the reduction depends on the degree of elevation of the arterial pressure. This proves that there is damaged response of the autonomic control to stress and in patients with *high normal pressure*. In [13] the authors have studied the autonomic balance of the heart through HRV in 11061 individuals for nine years. All hypertensive patients had reduced HRV already at the beginning of the study and this finding is valid for the entire range of hypertensive values of the arterial pressure. Among the normotensive individuals the low HRV values at the start of the study predict a higher risk of occurrence of hypertension for the 9-year period. HRV did not change considerably during the nine years in the normotensive individuals and in the hypertensive patients, although the difference between the variability in the two groups grew smaller with time. ***These results presuppose that the changes in the autonomic nervous system are included in the pathophysiology of the occurrence of hypertension, i.e., that they precede the onset of hypertension.***

CIRCADIAN NATURE OF THE AUTONOMIC BALANCE IN HYPERTENSION

The circadian variations of the heart rate and of the arterial pressure were described already back in the 17th century. In recent years, the development of the systems for 24-hour monitoring of arterial pressure has made it possible for us to distinguish between the circadian changes in hypertensive patients and in normotensive individuals. In normotensive individuals and in many of the patients with essential hypertension there is a nocturnal drop in the arterial pressure. With most forms of secondary hypertension, as well as sometimes with hypertensive disease, there is no such drop or even the nocturnal arterial pressure is higher than the daytime values. Verdecchia, Porcellati, Schilaci, et al. [14] have found that the risk of cardiovascular incidents is ten times higher in patients classified as non-dippers than in persons with white coat hypertension and in normotensive individuals. Shimada and Kario [15] associate the disordered circadian characteristic of arterial pressure with the damage to target organs. Mancia, Zanchetti, Agebiti-Rosei, et al. [16] claim that the 24-hour outpatient measurement of the arterial pressure is the best predictor for left-ventricular hypertrophy. In the past decade, the loss of the normal nocturnal fall in the arterial pressure has been associated with a rise in the incidence of cardiovascular incidents and damage of the heart, brain, vessels and kidneys.

The circadian characteristic of the arterial pressure is dependent on the biological clock (see *Chapter 4*). The link with the principal regulator of the circadian nature in the human organism – the hormone melatonin – is confirmed by the fact that melatonin intake reduces the nocturnal arterial pressure [17]. The circadian characteristics of the arterial pressure and of the heart rate are the best studied periodically changing indices of the cardiovascular system. In previous chapters it was pointed out that the stroke volume, the resistance of the peripheral vessels, the electrophysiological indices of cardiac activity, the plasma concentrations of the pressor hormones, the blood viscosity, coagulation and fibrinolysis also have circadian characteristics. These circadian characteristics are influenced by the circadian nature of the autonomic balance. As was noted already, the autonomic dysbalance apparently underlies the development of the hypertensive disease, whereas arterial hypertension is a major risk factor for the development of cardiovascular diseases. ***The disturbed circadian nature in the autonomic balance in arterial hypertension is a possible pathogenic cause for the circadian nature of cardiovascular incidents.*** Obviously, the study of the circadian characteristic of the autonomic balance in hypertensive patients is of great practical significance.

Guzzetti, Dassi, Pecis, et al. [18] have studied the circadian characteristic of the autonomic balance in normotensive individuals and in mildly hypertensive subjects through the frequency indices for HRV. Significant circadian changes in the two groups have been found for LF: during the nocturnal period the hypertensive patients had higher LF values. The difference between the daytime and the nocturnal values of LF in hypertensive patients decreases with the progression of hypertension. According to the authors, the reduced circadian nature of the sympathetic component in the balance is characteristic of the hypertensive patients. Chakko, Mulingtapang, Huicuri, et al. [19] have studied the time-domain and the frequency indices for HRV from 24-hour ECG recordings in normotensive individuals and hypertensive patients with left-ventricular hypertrophy. They have found that SDRR and MDRR are with lower values in hypertensive patients. The indice PNN50 follows the normal parasympathetic circadian nature in the controls – increasing during the night and decreasing during the day. In hypertensive patients this indice remains low throughout the entire 24-hour period. In normotensive individuals the values of LF are low during the night, rising in the morning and remaining high during the day. This circadian rhythm is absent in hypertensive patients. HF in hypertensive patients is characterized by lower values for a total of 21 hours from the entire 24-hour period, but the difference compared to the values in normotensive individuals is significant only for nine hours during the day and night. Kohara, Nishida, Maguchi, et al. [20] have studied on the basis of 24-hour ECG recordings the frequency indices from HRV in hypertensive patients without and with nocturnal fall in the arterial pressure. They have found that the non-dipper hypertensive patients have lower LF and HF values compared to the dipper hypertensive patients during the day. In the dipper patients LF has lower values during the night, and HF – significantly higher. These circadian fluctuations are strongly reduced in non-dipper hypertensive patients. The reduced circadian fluctuation of the autonomic balance with hypersympatheticotony and suppressed parasympathetic tone during the night explains also the absence of nocturnal fall in the arterial pressure. Damaged circadian fluctuation of the autonomic balance has also been found in [21]. The authors have studied normotensive individuals, hypertensive patients without LVH and hypertensive patients with LVH, and have confirmed that the nocturnal arterial pressure in normotensive individuals and in hypertensive patients without LVH is lower than the arterial pressure during the day. In the hypertensive patients with LVH there is no reduction of the nocturnal arterial pressure. Through normalized frequency indices of HRV it has been found that there is a decrease in LF from the daytime to the nocturnal period in normotensive individuals and in hypertensive patients without LVH. In hypertensive patients with LVH there are no changes between the daytime and the nocturnal values. What is more, in hypertensive patients with LVH, compared to normotensive individuals and hypertensive patients without LVH, the LF values increase during the night.

Apparently, the results of the cited studies do not give an idea about the circadian characteristic of the autonomic balance, differentiated according to the different degrees of the hypertensive disease: pre-hypertension, mild, moderate and severe. The above-mentioned hypothesis that the changes in the autonomic nervous system are involved in the pathophysiology of the occurrence of hypertension, i.e., that they precede the onset of hypertension, necessitate a more detailed study of the circadian nature of the autonomic balance in terms of differentiated severity of the disease. It can be expected that therapy aimed at normalization of the autonomic balance, including of its circadian characteristic, will influence the hypertensive disease effectively. As we emphasized earlier, such an approach is particularly expedient in the cases with mild hypertension (MH), when the therapeutic conduct also has a preventive role with respect to the development of more advanced stages of the hypertensive disease, leading to damage of target organs. With this motivation, we conducted a special comparative study of the circadian nature in the autonomic balance of the heart in mild hypertensive patients and normotensive individuals [22].

HEART AUTONOMIC BALANCE CHANGES IN MILDLY HYPERTENSIVE SUBJECTS

We studied 34 mild hypertensive patients without echocardiography (EchoCG) evidence of left-ventricular hypertrophy and without changes in the circadian characteristic of the arterial pressure according to data from its 24-hour monitoring. The group did not include persons with IHD (according to anamnestic data, ECG, and exercise ECG, if necessary), diabetes, chronic renal failure (CRF), neurological and other severe diseases. All persons were with sinus rhythm without conduction disorders. Seven patients had previous antihypertensive therapy, which was discontinued ten days before the studies. The values of the indices for HRV during rest, during the handgrip test and during the Valsalva manoeuvre, as well as the evaluations of TRABI for the indices, were compared with the respective data for 28 healthy subjects.

The values of the indices for HRV (with the exception of LF/HF) during rest in the morning and in the afternoon are significantly lower in hypertensive patients than the values in normotensive individuals. LF/HF in the hypertensive patients, albeit insignificant, is higher, due to the more substantial reduction of HF with respect to LF in these patients.

In mild hypertensive patients, as well as in healthy individuals during rest, there are no significant differences between the values of the indices in the different hours of the day (Tables 6.1 and 6.2).

In healthy subjects there are no significant differences between the values of the indices in the morning and in the afternoon during the handgrip test (Table 6.3). In the hypertensive patients there is a significant difference only for PNN50 (Table 6.4).

The comparison between the values of the indices in the morning in a study during rest and during the handgrip test in hypertensive patients indicates that the stimulation has no influence on the values of the indices from the time-domain area. In healthy individuals it is more pronounced for SDRR and MDRR. Apparently it is difficult to stimulate additionally the high tone of the sympathicus in the morning in the mild hypertensive patients. The values of the frequency indices are strongly reduced due to their high sensitivity to stimulation – both in healthy individuals and in mild hypertensive patients.

Table 6.1. Mean values of the HRV indices in the morning (M) and in the afternoon (AN) during rest in mildly hypertensive individuals.

	Mean(M)	SD(M)	Mean(AN)	SD(AN)	t value	p<
SDRR	31,50	10,0642	30,44	12,6496	0,3819	n.s.
MDRR	24,21	7,4500	23,62	9,8781	0,2772	n.s.
dSDRR	12,79	7,6308	13,53	10,4337	0,3317	n.s.
dMDRR	9,41	5,5548	10,09	8,4508	0,3900	n.s.
PNN50	2,29	3,7053	2,50	4,8508	0,1967	n.s.
RMSSD	19,68	10,9426	20,29	14,2521	0,2004	n.s.
VLF	124,12	76,9985	153,18	112,2822	1,2445	n.s.
LF	149,91	135,8489	138,68	125,1231	0,3547	n.s.
HF	66,68	57,3214	88,15	134,0900	0,8585	n.s.
LF/HF	3,22	2,6942	2,80	2,7509	0,6365	n.s.

Table 6.2. Mean values of the HRV indices in the morning (M) and in the afternoon (AN) during rest in normotensive individuals.

	Mean(M)	SD(M)	Mean(AN)	SD(AN)	t value	p<
SDRR	48,05	17,0279	44,8182	15,6985	0,6536	n.s.
MDRR	37,95	14,0898	35,3182	13,4780	0,6342	n.s.
dSDRR	20,73	12,6686	21,0000	10,0995	0,0790	n.s.
dMDRR	15,68	10,5216	15,9545	7,8829	0,0973	n.s.
PNN50	8,68	11,2224	9,6818	12,0806	0,2845	n.s.
RMSSD	32,59	21,0728	33,4091	17,1316	0,1413	n.s.
VLF	266,36	195,5480	219,4091	178,3762	0,8321	n.s.
LF	387,82	347,1007	376,0000	514,8662	0,0893	n.s.
HF	195,59	202,9252	205,9091	230,8042	0,1575	n.s.
LF/HF	2,64	1,8864	2,2800	1,5044	0,6954	n.s.

Table 6.3. Mean values of the HRV indices in the morning (M) and in the afternoon (AN) with the handgrip test in normotensive individuals.

	Mean(M)	SD(M)	Mean(AN)	SD(AN)	t value	p<
SDRR	39,68	20,1220	41,3182	20,4785	0,2673	n.s.
MDRR	31,77	16,4402	33,5455	16,8034	0,3537	n.s.
dSDRR	19,18	12,9270	18,6818	13,1017	0,1274	n.s.
dMDRR	14,73	10,3426	14,2727	9,8522	0,1493	n.s.
PNN50	6,77	9,7635	7,9545	12,7633	0,3450	n.s.
RMSSD	30,09	19,9521	29,8636	20,4806	0,0373	n.s.
VLF	88,68	94,2147	73,8182	64,7447	0,6099	n.s.
LF	65,45	65,0756	104,1818	173,7975	0,9788	n.s.
HF	44,27	52,5549	59,0455	98,1726	0,6222	n.s.
LF/HF	2,00	1,4616	3,0541	2,4394	1,7303	n.s.

Table 6.4. Mean values of the HRV indices in the morning (M) and in the afternoon (AN) with the handgrip test in mildly hypertensive individuals.

	Mean(M)	SD(M)	Mean(AN)	SD(AN)	t value	p<
SDRR	30,44	13,0481	26,53	11,1007	1,3314	n.s.
MDRR	24,24	10,2516	20,59	8,8937	1,5669	n.s.
dSDRR	14,85	19,2419	9,59	4,0160	1,5617	n.s.
dMDRR	9,03	5,9058	7,29	2,9080	1,5371	n.s.
PNN50	2,06	3,2747	0,71	1,2917	2,2410	0,05
RMSSD	18,12	10,8928	14,97	6,0477	1,4729	n.s.
VLF	55,71	43,8382	47,79	65,2978	0,5866	n.s.
LF	36,47	34,2329	47,15	70,7865	0,7917	n.s.
HF	18,79	22,4348	16,74	30,7131	0,3156	n.s.
LF/HF	3,35	3,3394	3,36	2,2157	0,0040	n.s.

In healthy individuals in the afternoon the handgrip test induces a slight change in the values of the HRV indices due to the stable tone of the sympathicus. In hypertensive patients the handgrip test causes a more considerable change, because the sympathicus has a lower tone compared to the morning and can be stimulated additionally. The tone of the sympathicus in the morning and in the afternoon differs more considerably in the hypertensive patients than in normotensive individuals.

The mean value of TRABI and the evaluations for the HRV indices in the comparative study between the reaction during the handgrip test and during rest in the morning and in the afternoon in mild hypertensive patients does not differ significantly from the mean value in healthy individuals (Figure 6.1). Nevertheless, for some of the indices – SDRR, MDRR, dSDRR, RMSSD – the values of TRABI are higher for the hypertensive patients. It can be considered that the sympathicus in the mild hypertensive patients has a more pronounced circadian nature compared to normotensive individuals. The indice PNN50 has lower values for TRABI in the hypertensive patients, which is probably due to the disordered circadian nature of the parasympathetic part of the balance, manifested even during the sympathetic test. The frequency indices have almost identical low values for the index in hypertensive patients and in normotensive individuals. The evaluation for LF/HF is higher in the normotensive individuals as a reflection of the preserved higher elasticity of the ratio between the sympathetic and the vagal component in the autonomic balance. In the mild hypertensive patients, due to the elevated sympathetic tone in the morning, the stimulation cannot induce changes to the same degree as in normotensive individuals, and the elasticity of the ratio is limited.

The comparison between the values of the indices for HRV during the Valsalva manoeuvre in the morning and in the afternoon showed no significant differences either in the healthy individuals (Table 6.5) or in the mild hypertensive patients (Table 6.6). The comparison between the values during rest and during the Valsalva manoeuvre in the morning and in the afternoon in hypertensive patients highlights the same reaction of the parasympathetic component in the autonomic balance. There is no marked circadian nature in the parasympathicotonia. Stimulation in the morning and in the afternoon did not reveal a difference in the values of the indices reflecting the vagal activity: PNN50, RMSSD and HF.

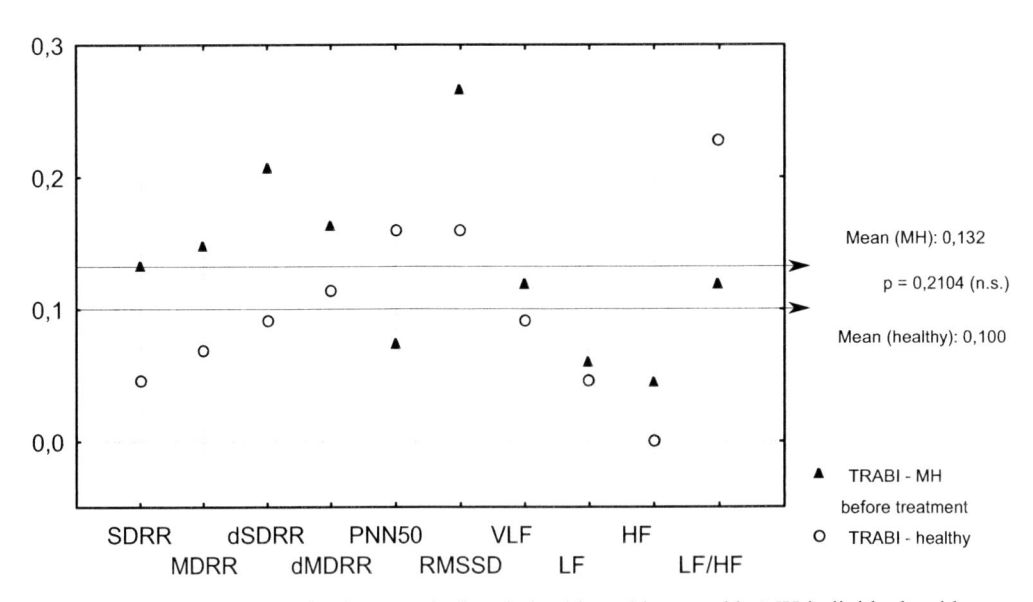

Figure 6.1. Values of TRABI for the HRV indices in healthy subjects and in MH individuals with comparison between morning and afternoon measurements during rest and with the handgrip test.

The mean value of TRABI and the evaluations for the HRV indices in the comparative study between the reaction during the Valsalva manoeuvre and during rest in healthy individuals is significantly higher than the mean value in the hypertensive patients (Figure 6.2). It can be claimed that the hypertensive patients have a reduced circadian nature of the parasympathetic part of the autonomic balance. All values of TRABI for the different indices are lower for the hypertensive patients. Only the evaluation for the LF index in the hypertensive patients has a higher value. This result is in support of the tendency in the sympathicotonia in hypertensive patients towards a more pronounced circadian nature, which was mentioned earlier.

Table 6.5. Mean values of the HRV indices in the morning (M) and in the afternoon (AN) with the Valsalva manoeuvre in normotensive individuals.

	Mean(M)	SD(M)	Mean(AN)	SD(AN)	t value	$p<$
SDRR	76,95	26,1506	76,55	37,8515	0,0417	n.s.
MDRR	60,32	20,5968	60,55	34,5691	0,0265	n.s.
dSDRR	31,86	13,1051	29,82	14,2182	0,4962	n.s.
dMDRR	22,73	9,4777	20,91	9,1230	0,6483	n.s.
PNN50	13,00	8,2865	12,41	10,0459	0,2128	n.s.
RMSSD	44,73	18,6807	41,77	18,9940	0,5208	n.s.
LF	131,14	96,3018	229,68	435,0151	1,0374	n.s.
HF	61,86	54,6838	49,05	60,4912	0,7373	n.s.
LF/HF	4,72	10,2313	8,52	19,4232	0,8119	n.s.

Table 6.6. Mean values of the HRV indices in the morning (M) and in the afternoon (AN) with the Valsalva manoeuvre in mildly hypertensive individuals.

	Mean(M)	SD(M)	Mean(AN)	SD(AN)	t value	p<
SDRR	58,3529	26,9769	67,0294	27,8573	1,3046	n.s.
MDRR	69,2941	134,5825	52,9412	25,1588	0,6964	n.s.
dSDRR	21,2941	15,3159	23,8529	13,2096	0,7377	n.s.
dMDRR	14,5588	9,7707	15,7353	8,8225	0,5211	n.s.
PNN50	6,0000	7,5438	6,3235	5,1211	0,2069	n.s.
RMSSD	29,5758	18,1625	31,5588	16,0155	0,4744	n.s.
LF	129,1471	230,9233	127,5294	147,9777	0,0344	n.s.
HF	26,8235	33,1027	38,6176	73,7625	0,8506	n.s.
LF/HF	5,7991	5,0378	7,9879	12,0451	0,9775	n.s.

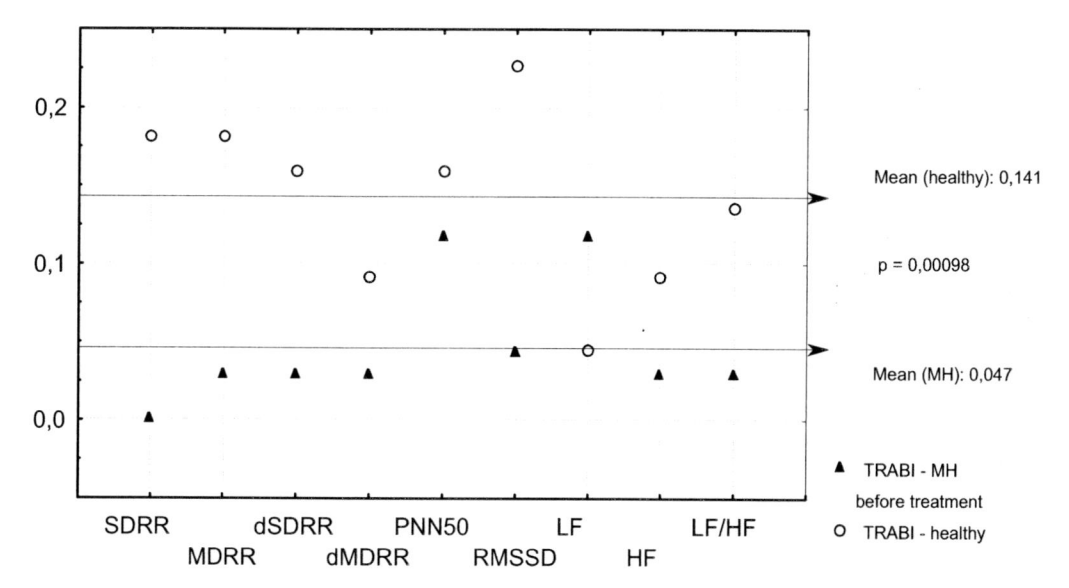

Figure 6.2. Values of TRABI for the HRV indices in healthy subjects and in MH individuals with comparison between morning and afternoon measurements during rest and with the Valsalva manoeuvre.

Several conclusions can be formulated as a summary of the above results. The values of the HRV indices in the mild hypertensive patients are significantly lower than the values in healthy subjects. *In the mild hypertensive patients there is a more pronounced circadian nature in the sympathetic component of the autonomic balance; the parasympathetic tone in them is suppressed; there is no circadian nature in the parasympathetic part of the autonomic balance.* This is also the main characteristic of the changes in the cardiac autonomic control in persons with mild hypertension.

ANTIHYPERTENSIVE THERAPY AND AUTONOMIC BALANCE

It was pointed out already that the hypersympatheticotony is connected with:

- development of left-ventricular hypertrophy and remodeling of the ventricles and of the vascular wall – autonomously and through activation of RAAS [23];
- endothelial dysfunction and the pace of atherosclerosis [24];
- development of the metabolic syndrome [25];
- occurrence of fatal ventricular rhythmic disorders [26].

The changes in the autonomic balance in hypertension, connected with the hypersympatheticotony and the reduced vagal tone, presuppose a therapeutic tactics that would not only impact the arterial pressure, but would also normalize the autonomic balance. Such a therapy would block the negative influence of hypersympatheticotony and would reduce the risk of cardiovascular incidents.

Studies have been conducted on almost all antihypertensive drugs to determine their influence on the autonomic balance. In [27] the study is on the influence of the beta-blockers and calcium antagonists on the autonomic balance in hypertensive patients through HRV. The authors indicate that with Atenolol therapy the values of the indices PNN50 and RMSSD, connected with the activity of the parasympathicus, increase, but there is no difference between the values of the indices SDRR and SDARR before and after therapy. There are no significant changes in the values of PNN50 and RMSSD before and after Carvedilol therapy. The values of SDNN decrease after Carvedilol therapy. The therapy with a long-active calcium antagonist (Lacidipine) does not provoke changes in the values of PNN50 and RMSSD. However, the therapy with calcium antagonists significantly increases the values of SDRR, which is associated with suppression of the sympathetic tone. This fact has also been found in another study, according to which administration of amplodipine reduces the urine and plasma level of noradrenalin [28].

In view of the link between the sympathetic and RAAS, studies have been conducted with angiotensin converting enzyme (ACE) inhibitors [29]. The authors have found that the therapy with ACE inhibitors improves the autonomic regulation of the heart. The angiotensin II-receptor blockers also influence the sympathetic nervous system. It is reported [30] that the presynaptic action of angiotensin on the sympathetic nerves increases the noradrenalin release. This explains the mechanism through which inhibition of RAAS can reduce the sympathetic nervous activity. The author has found a decrease of the sympathetic nervous activity after treatment with sartanes, determined using microneurography and the concentration of the plasma noradrenalin.

In the aspect of interest to us, the blockers of the imidazoline receptors are not very well studied. Nemes, Roth, Kapronczay, et al. [31] have demonstrated the effect of Rilmenidine on catecholamines and RAAS. Rilmenidine influences the activity of RAAS, which is evidenced by the reduction of the basal level of the plasma renin activity and of its level after standing up. At the same time, rilmenidine had no effect on the aldosterone level. The plasma level of catecholamines decreases significantly after rilmenidine. The level of the free radicals also decreases.

The cited survey allows the conclusion that modern antihypertensive drugs influence the autonomic balance to one degree or another. However, the evidence is controversial and it cannot be claimed with certainty that the beta-blockers, the calcium antagonists, the ACE inhibitors or the angiotensin II (AT II) receptor blockers restore the autonomic balance.

Due to the fact that the blockers of the imidazoline receptors act through central suppression of the sympathicus (in the brain stem), we undertook a study of the effect of such a blocker on the autonomic balance and its circadian nature [32]. As we found disorders in the autonomic balance in patients with mild hypertension in our earlier study [22] (see 6.3), we chose the same group of individuals. The choice was prompted by the consideration that if these drugs restore the autonomic balance, then their administration would result in a significant preventive effect – lower risk of changes in the target organs and cardiovascular incidents.

HEART AUTONOMIC BALANCE CHANGES AFTER CENTRAL SUPPRESSION OF THE SYMPATHETIC HYPERACTIVITY IN MILDLY HYPERTENSIVE SUBJECTS

We studied the influence of the blocker of the imidazoline receptors, Rilmenidine (Tenaxum®), on HRV and the circadian nature of the autonomic balance in the group consisting of the 34 patients for whom the autonomic balance was compared with the balance in healthy subjects (see 6.3).

Tenaxum® was administered for three months in a single dose of 1 mg in the morning. The results after the treatment are compared with the data before treatment and with the data in 28 normotensive individuals.

The values of the indices for HRV in the morning during rest after therapy with rilmenidine increase and there are no statistically significant differences between them and the values in healthy subjects (Tables 6.2 and 6.7). After the treatment, LF/HF has lower values both in the morning and in the afternoon compared to the values before the treatment (Tables 6.1 and 6.7), as a reflection of the attained reduction of the sympathicotonia. All time-domain indices in the treated mild hypertensive patients in the studies in the morning and in the afternoon failed to reach the values in healthy individuals, but exceeded the values before the treatment. Consequently, we can assume that HRV is restored and the autonomic balance normalized as a result of the therapy. At the same time, in spite of the therapy, the vagal tone in the mild hypertensive patients tends to be lower compared to normotensive individuals.

There are no significant differences between the values of the HRV indices during the handgrip test in the morning and in the afternoon after treatment (Table 6.8). This result is analogous to the result in healthy subjects (see Table 6.3). The significant difference between the values of PNN50 found before treatment is lacking. Apparently, the vagal tone is already stable during sympathetic stimulation.

Table 6.7. Mean values of the HRV indices in the morning (M) and in the afternoon (AN) during rest in mildly hypertensive individuals after treatment.

	Mean(M)	SD(M)	Mean(AN)	SD(AN)	t value	p<
SDRR	40,30	10,7295	36,40	13,3849	0,7189	n.s.
MDRR	31,40	7,7775	28,30	10,5520	0,7478	n.s.
dSDRR	18,30	11,0660	13,30	5,4171	1,2833	n.s.
dMDRR	13,80	9,0406	9,60	4,3512	1,3238	n.s.
PNN50	6,80	7,7431	2,50	2,7588	1,6543	n.s.
RMSSD	28,00	16,9575	20,60	8,8719	1,2227	n.s.
VLF	178,40	140,4779	180,00	160,2727	0,0237	n.s.
LF	253,50	206,8812	292,80	411,6910	0,2697	n.s.
HF	203,20	177,4466	127,10	131,7771	1,0888	n.s.
LF/HF	1,78	1,6489	2,68	1,6820	1,2150	n.s.

Table 6.8. Mean values of the HRV indices in the morning (M) and in the afternoon (AN) with the handgrip test in mildly hypertensive individuals after treatment.

	Mean(M)	SD(M)	Mean(AN)	SD(AN)	t value	p<
SDRR	36,10	12,6443	29,70	7,80385	1,3621	n.s.
MDRR	29,20	10,1412	23,30	6,76675	1,5304	n.s.
dSDRR	14,60	9,6056	10,50	4,03457	1,2445	n.s.
dMDRR	10,90	6,9992	7,70	3,05687	1,3249	n.s.
PNN50	3,50	4,3269	1,10	2,07900	1,5810	n.s.
RMSSD	21,60	12,9632	16,30	7,19645	1,1304	n.s.
VLF	82,90	85,9217	51,20	32,83562	1,0898	n.s.
LF	43,20	24,3985	58,30	52,20696	0,8286	n.s.
HF	56,20	97,6681	19,90	14,48716	1,1626	n.s.
LF/HF	3,53	4,6363	3,49	2,40874	0,0284	n.s.

Unlike the absence of reaction during the handgrip test in the morning with respect to the study during rest before therapy, after therapy the indices begin to respond to stimulation. This is yet another argument that reduction of the morning elevation of the sympathetic tone, found in hypertensive patients, has been achieved. The result of the comparison between the values of the indices during rest and during the handgrip test in the afternoon is similar. There is equalizing of the sympathetic tone in the different periods of the day and night, i.e., reduced circadian nature. Before treatment, the afternoon response to stress is considerably more strongly manifested compared to the one in the morning as a result of different sympathicotonia levels in the morning and in the afternoon.

After therapy, the values of TRABI for the HRV indices from the comparative study of the reaction during the handgrip test with respect to rest in the morning and in the afternoon decrease for almost all indices with respect to the values before treatment (Figure 6.3). The mean value from the evaluations of the index becomes equal to that in healthy individuals (Figure 6.4). This result is a new proof of the conclusion that the therapy suppresses the more pronounced circadian nature of the sympathetic component in the autonomic balance, which is specific for the mild hypertensive patients. However, for the frequency indices VLF, LF and HF the values of TRABI are higher compared to the values before the therapy, as well as to the values in healthy individuals. Apparently the therapy suppresses the hyperreaction of the sympathicus during stress and the frequency indices become more elastic when they reflect the time-related changes.

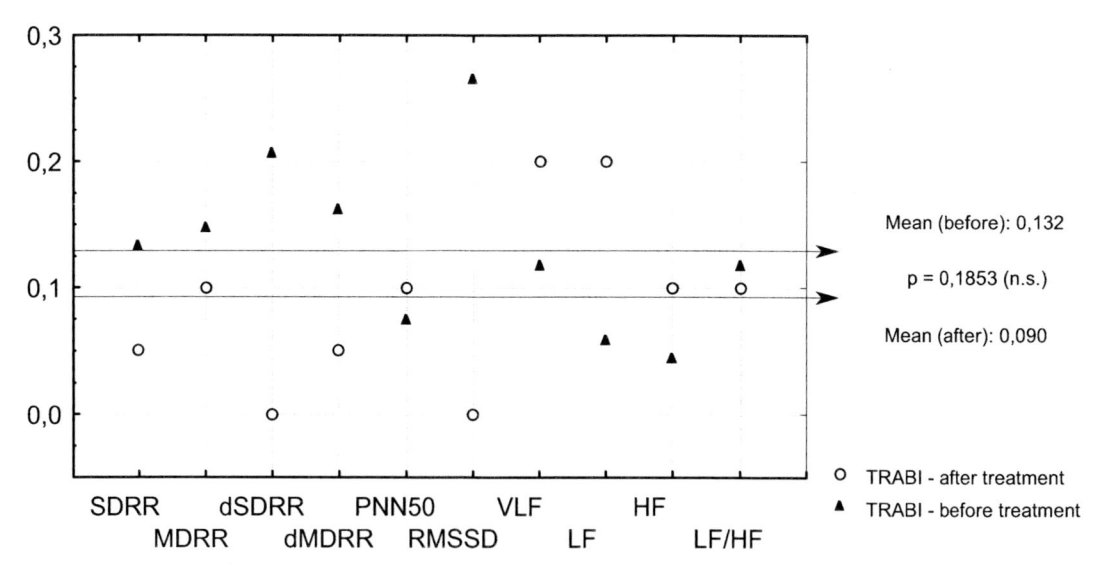

Figure 6.3. Values of TRABI for the HRV indices in MH individuals before and after treatment with comparison between morning and afternoon measurements during rest and with the handgrip test.

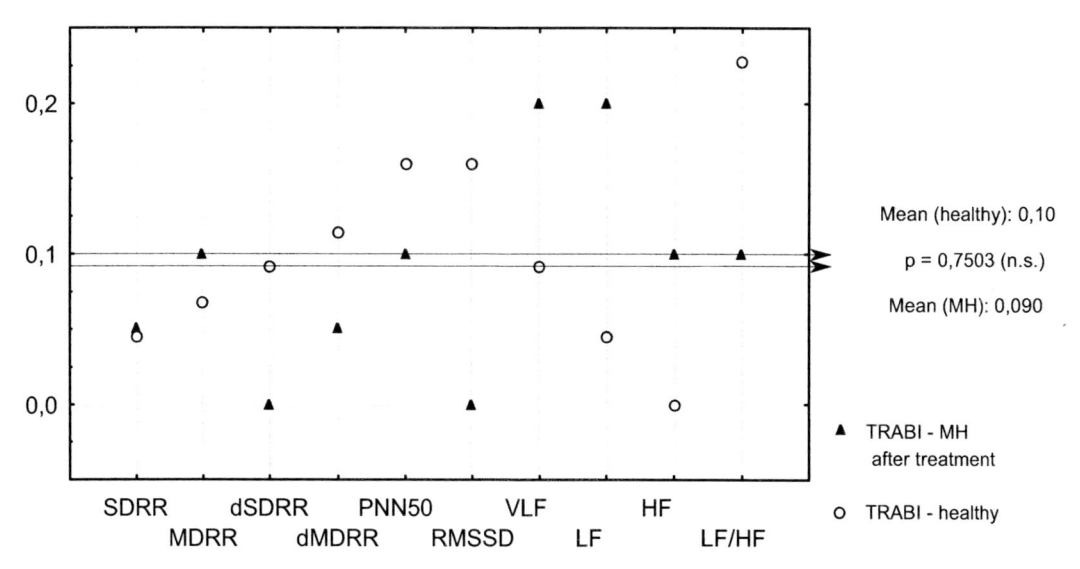

Figure 6.4. Values of TRABI for the HRV indices in MH individuals after treatment and in healthy subjects with comparison between morning and afternoon measurements during rest and with the handgrip test.

There are no significant differences between the values of the HRV indices during the Valsalva manoeuvre in the morning and in the afternoon both before and after the treatment of the mild hypertensive patients (Table 6.9). The same result is registered in healthy subjects as well (see Table 6.5). The values of the indices in the study in the morning indicate that the stimulation test induces a much stronger nervous reaction than the reaction before the therapy and even than the reaction in healthy subjects. There is lower vagal tone in the morning hours compared to the higher afternoon level when the reaction of the HRV indices to parasympathetic stimulation is weaker.

Table 6.9. Mean values of the HRV indices in the morning (M) and in the afternoon (AN) with the handgrip test in mildly hypertensive individuals after treatment.

	Mean(M)	SD(M)	Mean(AN)	SD(AN)	t value	$p<$
SDRR	96,20	41,7634	71,70	35,68	1,4104	n.s.
MDRR	77,70	37,5116	56,80	33,53	1,3136	n.s.
dSDRR	36,10	14,6625	25,30	11,35	1,8417	n.s.
dMDRR	24,40	8,3825	17,40	8,00	1,9100	n.s.
PNN50	12,00	6,3070	8,70	6,83	1,1223	n.s.
RMSSD	46,30	14,8776	35,10	14,70	1,6934	n.s.
LF	604,80	950,6274	205,40	238,62	1,2886	n.s.
HF	71,70	93,8735	51,60	77,89	0,5211	n.s.
LF/HF	22,16	43,0709	8,03	12,13	0,9985	n.s.

Compared to the values before the therapy, the values of TRABI for most HRV indices from the study of the reaction during the Valsalva manoeuvre with respect to rest in the morning and in the afternoon increase after the therapy (Figure 6.5). Before treatment, the mean value of TRABI for the HRV indices is significantly lower than the mean value in healthy subjects (see Figure 6.2). After treatment, the mean value of the index for the indices is practically identical to that in healthy subjects (Figure 6.6) and is significantly higher than the mean value of the index before treatment (Figure 6.5). This result shows restored circadian characteristic of the parasympathicus after treatment. For the indice PNN50 after treatment the value of TRABI is 0.350, being 0.118 before the treatment; for RMSSD the values for the index are lower than those in healthy individuals, but higher than the values before therapy. This is confirmation of the possibility to restore through the therapy the circadian characteristic of the parasympathetic component in the autonomic balance.

As a result, the therapy with Rilmenidine increases HRV, making it equal in the morning hours with the variability in healthy individuals. For most of the HRV indices this tendency is valid in the afternoon as well. *The therapy normalizes almost completely the autonomic balance in the mild hypertensive patients through the property of rilmenidine to restore the circadian nature of the parasympathicotonia with simultaneous suppression of the moderate morning hypersympatheticotony in the mild hypertensive patients.*

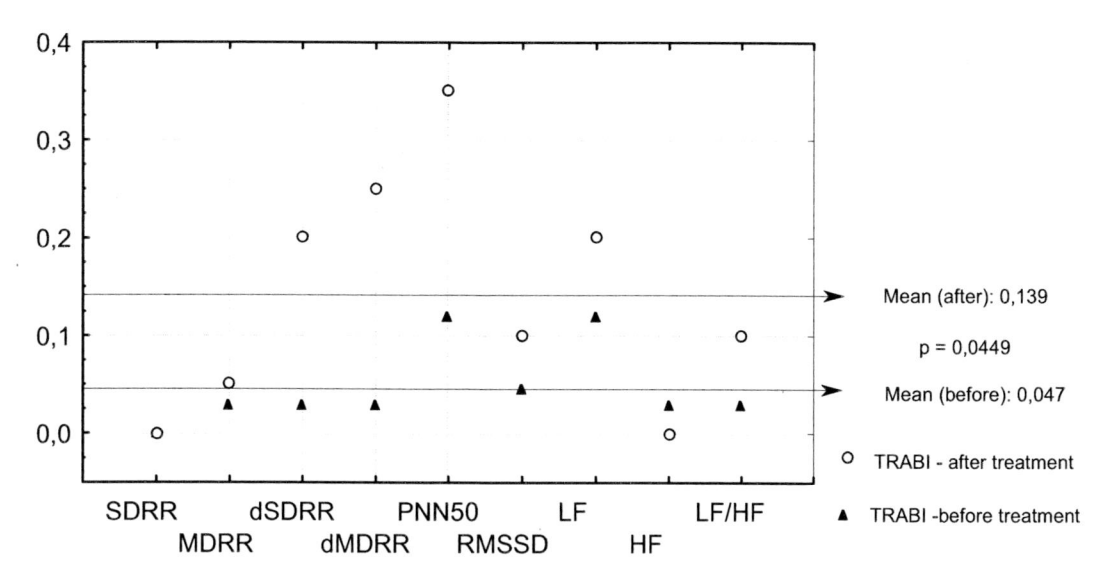

Figure 6.5. Values of TRABI for the HRV indices in MH individuals before and after treatment with comparison between morning and afternoon measurements during rest and with the Valsalva manoeuvre.

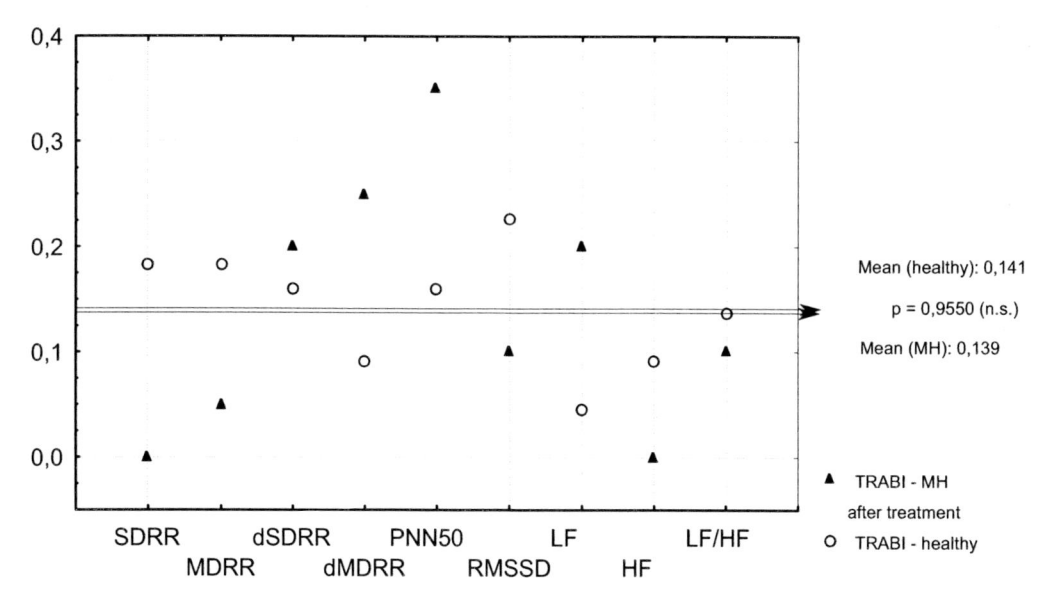

Figure 6.6. Values of TRABI for the HRV indices in MH individuals after treatment and in healthy subjects with comparison between morning and afternoon measurements during rest and with the Valsalva manoeuvre.

DIAGNOSTIC VALUE OF THE HRV INDICATORS FOR MILD HYPERTENSION

The results obtained in Section 6.3, revealing specific changes in the autonomic balance and in the circadian characteristics of the sympathetic and of its vagal component in mild hypertensive patients, raise the issue of the possibility to develop with the indices for HRV an algorithm for detecting persons who need to start antihypertensive therapy early in spite of the borderline or even normal values of their arterial pressure. Already back in 2002, we formulated the hypothesis that the criterion for classifying a concrete subject with high normal values of the arterial pressure in the group of mild hypertensive patients or in the population of healthy individuals is whether stable changes have occurred in the circadian nature of the autonomic balance [33]. This hypothesis is in compliance with the results of large-scale studies on the predictive value of the lowered HRV for detecting future hypertension. The results of the study conducted by us to discriminate between healthy subjects (normotensive individuals) and mild hypertensive patients on the basis of data from the HRV indices and taking into account the circadian specificities of the autonomic balance in the two groups of persons are extremely encouraging. The practical result of the study is a convenient rule for clinical practice to decide on the classifying of a concrete individual in one of the two groups.

Two groups of individuals were included in the study - normotensive subjects (n=28) and patients with mild hypertension (n=23). The gender distribution was 14 men and 14 women in the normotensive group, and 7 men and 15 women with mild hypertension. The mean age in the groups was 50.8, range 21–76 and 50.1, range 37–75 respectively.

A classification model was developed and first probated on an independent set of 22 mild hypertension patients (8 men and 14 women, mean age 49.3, range 26–63) and 16 normotensive individuals (8 men and 8 women, mean age 41, range 35-60).

The following criteria were used for the groups of normotensive and mildly hypertensive individuals:

Criteria	Mild Hypertension	Normotensive
Arterial pressure* (mm Hg):	95<DAP<114 140<SAP<159	DAP< 95 SAP<140
24-hour observation:	normal night dip >10% (the absence of a dip during the night correlates to the existence of LVH)	normal arterial pressure values
ECG and EchoCG:	LVH signs absent	
Absence of any indications for:	– severe hypertension; – cardiac insufficiency; – myocardial infarction; – diabetes; – neurological diseases; – renal failure; – other system diseases	
Cancellation of medication:	– 5 days prior to examination, in the cases of treatment with beta-blockers; – 3 days prior to examination, in the cases of treatment with calcium antagonists, diuretics, ACE inhibitors, or centrally acting drugs	

* Diastolic (DAP) and systolic (SAP) arterial blood pressure (ABP) values in this study were determined on the basis of 24-hour measurements.

The above-described indicators were obtained three times for each individual of both groups, from the recordings of R-R intervals in resting state, handgrip test and Valsalva manoeuvre. The indicators pertaining to each specific test were marked and an index was added to the abbreviation: _1 (resting state), _2 (handgrip test), _3 (Valsalva manoeuvre). The R-R intervals recordings in the three tests were taken separately in the morning (8-9 a.m.) and in the afternoon (2-3 p.m.). The HRV indicators used are characterized by a wide range of values, due to the different measurement units. For example, the values of the LF band spectral area could exceed 300, whereas the ratio LF/HF is in the range of one. In order to introduce an invariant scale for the measurement units, we standardized the values of the indicators with respect to the so-called common (intergroup) mean value. The common mean value of the p-th indicator is calculated using the following formula:

$$CMI_p = (N_{hyp} \cdot MI_{p\_hyp} + N_{norm} \cdot MI_{p\_norm})/(N_{hyp} + N_{norm}),$$

where N_{hyp} and N_{norm} are the number of hypertensive and normotensive subjects, and MI_{p_hy} and MI_{p_norm} are the mean values of the p-th indice of hypertensive and normotensive subjects, respectively. The standardization of the individual value for each indice is done by dividing it by its corresponding common mean value.

In order to synthesize a reliable, statistically significant, very specific and highly sensitive algorithm for optimal discrimination between normotensive and mildly hypertensive subjects, we used a stepwise linear discriminant analysis (LDA), thoroughly described in a number of manuals on multivariate statistical analysis [34,35] and in studies on the application of statistical methods in medicine [36, 37, 38]. The choice of this classification method was based on its adequacy to the study object. The discrimination analysis is a fundamental algorithm in evaluation of factors maximizing inter-group dispersion. In addition, it corresponds to the statistical characteristics of the data. They fall in the limits of the linear theory for a class of logistic distributions generating a linear discriminator [34, 35].

The practical result from the application of LDA is a synthesis of a set of indicators (considerably fewer as a rule, compared to the full set of indicators available), possessing optimal power for correct classification of the subjects of both groups. The LDA algorithm parametrizes two linear discriminant functions (hyper-planes), D_1 and D_2, running through the centres of the multidimensionally-distributed data for the two groups of subjects:

$$D_1 = a_{10} + a_{11}I_1 + a_{12}I_2 + \ldots + a_{1k}I_k \quad \text{and} \quad D_2 = a_{20} + a_{21}I_1 + a_{22}I_2 + \ldots + a_{2k}I_k.$$

Here I_p (p=1, 2, ..., k) are the indices used, a_{jp} (j=1,2; p=0,1,...,k) are coefficients whose estimates were determined on the basis of the two training sets. When the indices of a given subject are entered in D_1 and D_2, if $D_1 > D_2$, he is associated to the first group and the opposite. Obviously, the algorithm can be reduced to a single substitution, by the operation

$$D_1 - D_2 = (a_{10} - a_{20}) + (a_{11}-a_{21})I_1 + (a_{12}-a_{22})I_2 + \ldots + (a_{1k}-a_{2k})I_k.$$

Then the comparison $D_1 \gtrless D_2$ is equivalent to

$$D = (a_{10} - a_{20}) + (a_{11}-a_{21})I_1 + (a_{12}-a_{22})I_2 + \ldots + (a_{1k}-a_{2k})I_k \gtrless 0,$$

or

$$D = a_1I_1 + a_2I_2 + \ldots + a_kI_k \gtrless L,$$

where $a_p = (a_{1p} - a_{2p})$, $(p=1,2,\ldots,k)$ and $L = (a_{20} - a_{10})$ is the so-called limit value.

In the case of classifying a subject to the normotensive or hypertensive group, the values of the constellation indices I_p (p=1, 2, ... k) are multiplied by the coefficients a_p and the obtained are summed. The sum D is compared to the limit value L. If D>L, the individual is classified as hypertensive, and the opposite.

Morning Studies

The stepwise LDA procedure leads to the synthesis of a linear discriminator of very high statistical significance. The value 5.29 of the calculated F-criterion rejects the hypothesis of identity (non-discrimination) of the two groups of subjects with a confidence interval p<0.0003. The statistical indicators of the linear discriminator are listed in table 6.10.

Table 6.11 lists the coefficients by which each indice participating in the linear discriminator is multiplied.

The linear discriminator (LD), applied to the data of each individual from both groups, yields a very high percentage of correct classification - the weighted percentage of correct classification of the individuals of both groups is 92%. One individual only out of 23 with mild hypertension was incorrectly classified as normotensive, and out of 28 normotensive, just 4 were misclassified (table 6.12).

Table 6.10. Discriminant Analysis Summary.

STAT. DISCRIM. ANALYSIS N=51	Discriminant Function Analysis No of vars in model: 9; Grouping: GROUP (2 grps) Wilks' Lambda: .36173; approx. F =5.2935; p< .0003					
	WILKS_L	PARTIAL_L	F-REMOVE	P_LEVEL	TOLER.	1-TOLER.
MDRR_1	0.413735	0.874300	3.881838	0.059146	0.335035	0.664965
VLF_1	0.446797	0.809604	6.349638	0.017961	0.550363	0.449637
LF_1	0.428116	0.844931	4.955290	0.034551	0.273988	0.726012
RRA_2	0.396509	0.912284	2.596045	0.118761	0.503226	0.496774
VLF_2	0.463077	0.781140	7.564853	0.010493	0.418065	0.581936
MDRR_3	0.432938	0.835521	5.315174	0.029057	0.391256	0.608744
PNN50_3	0.385663	0.937939	1.786525	0.192505	0.550535	0.449465
LF_3	0.550250	0.657389	14.071580	0.000851	0.314934	0.685066
LF_HF_3	0.442955	0.816626	6.062866	0.020476	0.596204	0.403796

Table 6.11. Linear discriminator, based on the data from the morning studies.

Indicator (I_p, p=1, 2,...,9)	Coefficient (a_p, p=1, 2,...,9)
MDRR_1	-54
VLF_1	-21
LF_1	23
RRA_2	123
VLF_2	23
MDRR_3	40
PNN50_3	-10
LF_3	-33
LF/HF_3	21
Limit value (L):	**100**

Table 6.12. Classification ability of the linear discriminator.

Group	Correct classification	Correct classification %	Incorrect classification	Incorrect classification %
Mild hypertension	22	95.7	f.n.**: 1	f.n.: 4.3
Normotensive	24	85.7	f.p.**: 4	f.p.: 14.3
Total	46	91.9*	5	8.1*

* Common (intergroup) mean percentage; ** f.n. – false negative, f.p. – false positive.

Figure 6.7 shows the profiles of the two groups, based on the values of the linear discriminator for each individual.

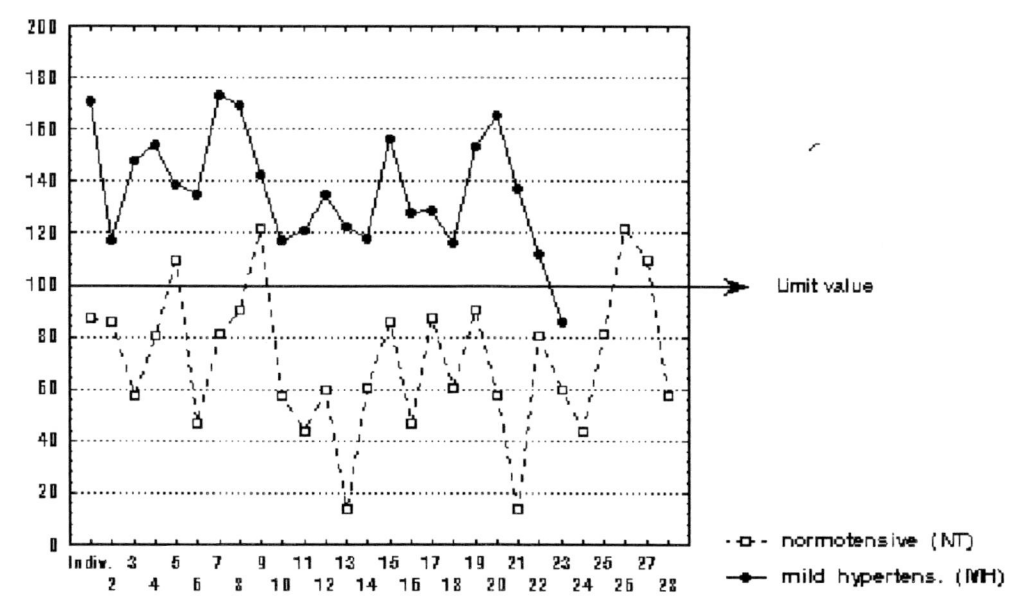

Figure 6.7. Distribution of the values of the linear discriminator for each individual of the MH and NT cases - morning studies.

Afternoon Studies

The stepwise LDA procedure leads to the synthesis of a linear discriminator with a high statistical significance. The value 3.114 of the calculated F-criterion rejects the hypothesis for identity (non-discrimination) of the two groups of subjects, with a confidence interval $p<0.086$. The statistical indicators of the discriminator are listed in table 6.13.

Table 6.13. Discriminant Analysis Summary

STAT. DISCRIM. ANALYSIS	Discriminant Function Analysis No of vars in model: 9; Grouping: GROUP (2 grps) Wilks' Lambda: .39105; approx. F =3.114; p< .0086					
N=51	WILKS_L	PARTIAL_L	F-REMOVE	P_LEVEL	TOLER.	1-TOLER.
RRA_1	0,423424	0,923548	1,986743	0,171509	0,111342	0,888658
SDRR_1	0,403828	0,968362	0,784121	0,38467	0,006713	0,993287
MDRR_1	0,40477	0,966109	0,841908	0,367981	0,006097	0,993903
LF_1	0,430192	0,909018	2,402126	0,134258	0,054139	0,945861
SDRR_2	0,48944	0,798978	6,038372	0,021606	0,00503	0,99497
MDRR_2	0,496527	0,787575	6,473293	0,017804	0,004865	0,995136
MED_2	0,467287	0,836857	4,678742	0,040719	0,141364	0,858636
LF_2	0,445073	0,878625	3,315399	0,081125	0,060754	0,939246
dRRA_3	0,461361	0,847605	4,315081	0,048646	0,025883	0,974117
dSDRR_3	0,409521	0,954901	1,133503	0,297624	0,051144	0,948856
dMDRR_3	0,449602	0,869773	3,593413	0,070114	0,024182	0,975818
PNN50_3	0,409614	0,954684	1,139219	0,296437	0,065806	0,934194

Table 6.14 lists the coefficients by which each indice participating in the linear discriminator is multiplied.

Table 6.14. Linear discriminator, based on the data from the afternoon studies.

Indicator (I_p, p=1,2,...,12)	Coefficient (a_p, p=1,2,...,12)
RRA_1	-66.5
SDRR_1	46.5
MDRR_1	-48.0
LF_1	-7.0
SDRR_2	121.0
MDRR_2	-122.0
MED_2	85.0
LF_2	7.0
dRRA_3	-45.5
dSDRR_3	-17.0
dMDRR_3	39.0
PNN50_3	8.0
Limit value (L):	1

The linear discriminator, applied to the data of each individual from both groups of patients, provides very high percentage of correct classification - the weighted percentage of correct classification of the individuals of both groups is 94.2%. One individual only, out of

23 with hypertension, was incorrectly classified as normotensive. Out of 28 normotensive subjects, just 2 were classified as hypertensive (table 6.15).

Table 6.15. Classification ability of the linear discriminator.

Group	Correct classification	Correct classification %	Incorrect classification	Incorrect classification %
Mild hypertension	22	95.7	f.n.**: 1	f.n.: 4.3
Normotensive	26	92.9	f.p.**: 2	f.p.: 7.1
Total	48	94.2*	3	5.8*

* Common (intergroup) mean percentage; ** f.n. – false negative, f.p. – false positive.

Figure 6.8 shows the profiles of the two groups, based on the values of the linear discriminator for each individual.

Testing the Model on Independent Samples

The discrimination power of the synthesized models was tested on two groups of subjects, which were not included in the training sets. The classification results, from the corresponding morning and afternoon examinations, are shown in table 6.16 and table 6.17.

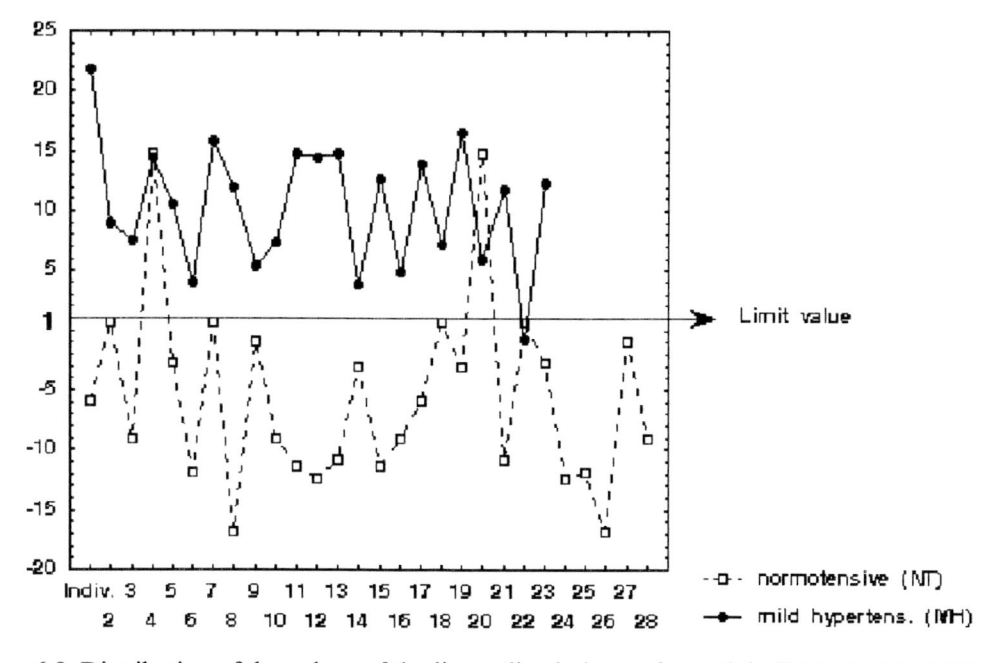

Figure 6.8. Distribution of the values of the linear discriminator for each individual of the MH and NT cases - afternoon studies.

Table 6.16. Classification ability of the linear discriminator - morning studies.

Group	Correct classification	Correct classification %	Incorrect classification	Incorrect classification %
Mild hypertension	22	100.0	f.n.**: 0	f.n.: 0.0
Normotensive	14	87.5	f.p.**: 2	f.p.: 12.5
Total	36	94.7*	2	5.3*

* Common (intergroup) mean percentage; ** f.n. – false negative, f.p. – false positive.

Table 6.17. Classification ability of the linear discriminator - afternoon studies.

Group	Correct classification	Correct classification %	Incorrect classification	Incorrect classification %
Mild hypertension	16	72.7	f.n.**: 6	f.n.: 27.3
Normotensive	12	75.0	f.p.**: 4	f.p.: 25.0
Total	28	73.7*	10	26.3*

* Common (intergroup) mean percentage; ** f.n. – false negative, f.p. – false positive.

Results: *Morning Studies*

The linear discriminator includes indicators from all three tests: resting state (MDRR_1, VLF_1, LF_1), handgrip test (RRA_2, VLF_2), and Valsalva manoeuvre (MDRR_3, PNN50_3, LF/HF_3). It can be presumed that the participation of provocative tests in the LD with a considerable number of indicators is necessitated by their power to detect even slight changes in vegetative control, specific for the cases of mild hypertension.

The spectral indicators VLF and LF in the LD reflect a reduced spectral area in mild hypertensive patients, compared to normotensive subjects. Due to the same trend, observed in the HF indice, it can be concluded that in the mild hypertensive cases an overall decrease of the spectral area is evident, showing lowered heart rate variability. Similar evidence can be found in the time-domain indicators MDRR and PNN50. Their values are lower in the mild hypertension cases compared to the normotensive. The predominant depression of the overall spectral area, observed in the mild hypertensive cases, leads to a hypothesis of change in the vegetative nervous system balance towards hypersympatheticotony and reduced parasympathetic tone, even in cases with mild arterial hypertension.

The presence of the VLF indice in the LD, based on data from the resting state and handgrip tests, can also be viewed as an early prognostic factor of imminent LVH, since it is connected to the activity of the renin-angiotensin system.

The Valsalva manoeuvre stimulates the parasympathetic part of the vegetative nervous system. In this aspect, the inclusion of PNN50 and LF/HF indicators in the LD relates to a disturbance in the vegetative control with suppressed vagal stimulation. In this case the frequency ratio LF/HF shows considerable difference in the HF spectral range between hypertensive and normotensive cases. The spectral area HF is 2.7 times smaller in the cases of hypertension.

Afternoon Studies

An equal number of indicators from the three tests are included in the linear discriminator: resting state (RRA_1, SDRR_1, MDRR_1, and LF_1), handgrip test (SDRR_2, MDRR_2, MED_2, and LF_2), and Valsalva manoeuvre (dRRA_3, dSDRR_3, dMDRR_3, PNN50_3). The capability of the provocative tests to detect slight changes in the vegetative control, specific for mild hypertension, is confirmed.

The time-domain indicators (a total of 10) in the linear discriminator prevail over the ones derived from the frequency domain. This results from the connection between the time-domain indicators and the changes occurring in the HRV in patients with mild hypertension. The absence of frequency variables reflects the trend toward stabilization of the vegetative nervous system during afternoon hours. The only frequency indice present is LF in two of the tests – resting state and handgrip. Its inclusion was necessitated by the hypersympatheticotony of mild hypertensive patients.

The only participation of tachogram indicators - the differences between adjacent RR-intervals (dRRA_3, dSDRR_3, and dMDRR_3) in the Valsalva manoeuvre, is a result from the stimulation of the parasympathetic system during this test. Its short duration acts predominantly on the RR-variation of adjacent intervals.

The analysis of the results obtained with the synthesized models for classification of mild hypertensive and normotensive individuals, from the training sets and the independent groups, allows to recommend for practical application the linear discriminator obtained from data of the morning studies. The main reason was already pointed-out: the tendency of stabilisation of the vegetative nervous system in the afternoon hours, which masks the HRV changes characteristic for mild hypertension. The considerable difference between the limit values in the two models – respectively 100 for the morning examinations and 1 for the afternoon ones, is a proof for the markedly reduced distance between the centers of the multidimensional data distributions of the normotensive and mild hypertensive cases in the afternoon hours. There is a tendency of overlapping the two groups distributions. This deduction was confirmed by the data of table 16 and table 17, showing maintaining the high discrimination power of the morning model and the lower discrimination of the afternoon one. An additional consideration in favour of the use of the morning model is the well-known fact, that the risk of cardiovascular complications is higher in the morning hours.

The results of discrimination between groups of patients with mild hypertension and normotensive subjects provide convincing evidence of the possibility for correct classification of individuals in the two groups. The adequacy of the discriminant models (91.9% and 94.2%) is illustrated by the specific participation of indicators from the time and frequency domains as a result of their capacity to reflect the vegetative nervous system balance.

REFERENCES

[1] Singh, JP; Larson, MG; Tsuji, H; Evans, JC; O Donel, CJ; Levy D. Reduced Heart Rate Variability and New-Onset Hypertension. *Hypertension*, 1998, 32, 293-297.

[2] Chobanian, A; Barkis, GL; Black, HR; et al. The seventh report of the Joint National Committee on Prevention, Detection, Evaluation, and Treatment of High Blood Pressure. *JAMA*, 2003, 289, 2560-2582.

[3] Perin, PC; Maule, S; Quadri, R. Sympathetic nervous system, diabetes, and hypertension. *Clin Exp Hypertens*, 2001, 23, 45-55.

[4] Rabbia, F; Martini, G; Cat Genova, G; Milan, A; Chiandussi, L; Veglio, F. Antihypertensive drugs and sympathetic nervous system. *Clin Exp Hypertens*, 2001, 23, 101-111.

[5] Yasuko, N. Heart rate Variability in Stroke Patients: Relationship between Blood Pressure Variability and Arteriosclerosis. *Jpn J Rehabil Med*, 2002, 39, 785-792.

[6] Anderson, ED; Sinkey, CA; Lawton, WJ; Mark, AL;. Elevated sympathetic nerve activity in borderline hypertensive humans. *Hypertension*, 1989, 14, 177-183.

[7] Guzzetti, S; Piccaluga, E; Casati, R; Cerutti, S; Lombardi, F; Pagani, M; Malliani, A. Sympathetic predominance in essential hypertension: a study employing spectral analysis of heart rate variability. *J Hypertens*, 1988, 6, 711-717.

[8] Petretta, M; Bianchi, V; Marchiano, F; Themistoclakis, S; Canonico, V; Sarno, D; Lovino, G; Bonaduce, D. Influence of left ventricular hypertrophy on heart period variability in patients with essential. *J Hypertens*, 1995, 13, 1299-1306.

[9] Paccirillo, G; Munizzi, M; Fimognari, F; Margliano, V. Heart rate variability in hypertensive subjects. *Intern J Cardiol*, 1996, 53, 291-298.

[10] Mussalo, H; Vanninen, E; Ikaheimo, R; Laitinen, T; Laakso, M; Lansimies, E; Hartikainen, J. Heart rate variability and its determinants in patients with severe or mild hypertension. *Clin Rhysiol*, 2001, 21, 594-604.

[11] Martini, G; Rabbia, F; Gastaldi, L; Riva, P; Sibona, MP; Morra di Cella, S; Chiandussi, L; Veglio, F. Heart rate variability and left ventricular diastolic function in patients with borderline hypertension with and wihout left ventricular hypertrophy. *Clin Exp Hypertens*, 2001, 23, 77-87.

[12] Lucini, D; Mela, GS; Malliani, A; Pagani, M. Impairment in Cardiac Autonomic Regulation Preceding Arterial Hypertension in Humans. *Circulation*, 2002, 106, 2673-2679.

[13] Schoeder, E; Liao, D; Chambless, LE; Prineas, RJ; Evans, GW; Heiss, G. Hypertension, Blood Pressure, and Heart Rate Variability. Atherosclerosis Risk in Communities Study (ARIC). *Am J Epidemiol*, 1989, 129(4), 687-702.

[14] Verdecchia, P; Porcellati, C; Schilaci, G; Borgioni, C; Ciucci, A; Battistelli, M; Guerrieri, M; Gatteschi, C; Zampi, L; Santucci, C; Reboldi, G. Ambulatory blood pressure an independent predictor of prognosis in essential hypertension. *Hypertension*, 1994, 24, 793-801.

[15] Shimada K, Kario K. Altered circadian rhythm of blood pressure and cerebrovascular damage. *Blood Press Monit*, 1997, 2, 333-338.

[16] Mancia, G; Zanchetti, A; Agebiti-Rosei, E; Benemio, G; De Cesaris, R; Fogari, R; Pessino, AC; Porcella.ti, C; Salvetti, A; Trimarco, B; for the SAMPLE Study Group. Ambulatory blood pressure is superior to clinic blood pressure in predicting treatment-induced regression of left ventricular hypertrophy. *Circulation*, 1997, 95, 1464-1470.

[17] Scheer, FJ; Van Montfrans, GA; Van Someren, EJ; Miruhu, G; Buijs, RM. Daily Nighttime Melatonin Reduces Blood Pressure in Male Patients With Essential Hypertension. *Hypertension*, 2004, 43, 192-197.

[18] Guzzetti, S; Dassi, S; Pecis, MJ; Casati, R; Masu, AM; Longoni, P; Tinelli, M; Cerutti, S; Pagani, M; Malliani, A. Altered pattern of circadian neural control of heart period in mild hypertension. *J Hypertens*, 1991, 9, 831-838.

[19] Chakko, S; Mulingtapang, RF; Huicuri, HV; Kessler, KM; Materson, BJ; Myerburg, RJ. Alterations in heart rate variability and its circadian rhythm in hypertensive patients with left ventricular hypertrophy free of coronary artery desease. *Am Heart J*, 1993,126, 1364-1372.

[20] Kohara, K; Nishida, W; Maguchi, M; Hiwada, K. Autonomic nervous function in non-dipper essential hypertensive subjects. Evaluation by power spectral analysis of heart rate variability. *Hypertension*, 1995, 26, 808-814.

[21] Muiesan, ML; Rizzoni, D; Zulli, R; Castellano, G; Porteri, BE; Agebiti-Rossei, E. Circadian changes of power spectral analysis of heart rate in hypertensive patients with left ventricular hypertrophy. *High Blood Press*, 1996, 5, 166-172.

[22] Prokopova, R; Matveev, M. Heart autonomic balance changes in mildly hypertensive subjects. Method of assessment, characteristics, treatment. *J Hypertens*, 2004, 2, Suppl 2, S209-S210.

[23] EL-Gharbawy, AH; Nadig,VS; Kotchen, JM; Grim, CE; Sagar, KB; Kaldunski, M; Hamet, P; Pausova, Z; Gaudet, D; Gossard, F; Kotchen, TA. Arterial pressure, left ventricular mass, and aldosterone in essential hypertension. *Hypertension*, 2002, 37, 845-850.

[24] Valensi, P; Verdier. Diabetes: Challenges of the Millenium Risk with Moxonidine. 2004. Available from: *http://webcasts.prous.com/esc2004%5Fsolvay%5Fmoxo/program.asp.*

[25] Heinz, R; Bernhard, M. Abdominal obesity and elevated activity of the sympathetic nervous system. *Herz*, 2003, 28, 668-673.

[26] Mandawat, MK; Wallbridge, DR; Pringle, SD; Riyami, AS; Latif, S; Macfarlane, PW; Lorimer, AR; Cobbe, SM. Impaired Heart Rate Variability and Increased Ventricular Ectopic Activity in Patients With Left Ventricular Hypertrophy. *J of Electrocardiol*, 1994, 27, Suppl, 179-181.

[27] Rabbia, F; Martini, G; Sibona, M; Grosso, T; Simondi, F; Chiandussi, L; Veglio, F. Assessment of Heart Rate Variability After Calcium Antagonist and Beta-Blocker Therapy in Patients with Essential Hypertension. *Clin Drug Invest*, 1999, 17, 111-118.

[28] Hamada, T; Watanabe, M; Kaneda, TJ; et al. Evaluation of changes in sympathetic nerve activity and heart rate in essential hypertensive patients induced by amlodipine and nifedipine. *J Hypertens*, 1998, 16, 111-118.

[29] Kamenskaya, E; Stepanov, A. Enalapril and Heart Rate Variability in Patients with Arterial Hypertension. 2000. Available from: *http://www.fac.org.ar/cvirtual/tlibres/tnn2396i/tnn2396.htm.*

[30] Esler, M. Differentiation in the effects of angiotensin II receptor blocker class on autonomic function. *J Hypertens*, 2002, 20, Suppl 5, S13-S19.

[31] Nemes, J; Roth, E; Kapronczay, O; Mozsik, G. Effect of rilmenidine , a centrally acting imidazoline agonist on the renin-angiotensin-aldosterone and catecholamine system and on the indices indicating oxidative stress in patients with essential hypertension. *Hypertension*, 2000, 18, Suppl.2, S119-S119.

[32] Prokopova, R; Matveev, M; Nachev, Ch. J. Heart autonomic balance changes after peripheral and central suppression of sympathetic hyperactivity in mildly hypertensive individuals. *Hypertens*, 2005; 23, Suppl 2, S382.
[33] Matveev, M; Prokopova, R. Diagnostic value of the RR-variability indicators for mild hypertension. *Physiol. Meas.*, 2002, 23, 671-682.
[34] Geisser, S. Predicative discrimination. Multivariate Analysis. In: Krishnaiah P, editor. *Discrimination and Cluster*. New York: Academic Press; 1966, 149-163.
[35] Enis, P; Geisser, S. Sample discriminants which minimize posterior squared error loss. *South Afr. Statist. J.*, 1970, 4, 85-93.
[36] Rao, CR. Advanced *Statistical Methods in Biometric Research*. New York:Wiley; 1952.
[37] Lusted, L. Introduction *to Medical Decision Making*. Springfield, USA: Charles C. Thomas; 1985.
[38] Van Bemmel, JH. *Medical Decision Making*. Amsterdam: North-Holland; 1985.

Chapter 7

MODELS OF THE AUTONOMIC DYSBALANCE IN ISCHAEMIC HEART DISEASE

Rada Prokopova
St. Anna University Hospital, Sofia, Bulgaria
Mikhail Matveev
Centre of Biomedical Engineering, Bulgarian Academy of Sciences, Bulgaria

CHANGES IN THE AUTONOMIC BALANCE IN CORONARY ARTERY DISEASE

The ischaemic heart disease occupies a leading position among the causes for hospitalization and mortality in recent years. In spite of the development of pharmacological therapy, the methods of invasive cardiology and cardiosurgery, the mortality and the invalidization after IHD remain high. This requires more efforts for identifying the factors that trigger the appearance and development of IHD. As was indicated in the previous chapters, the disturbed autonomic balance is connected with one of the main risk factors for IHD, namely hypertension. In recent years, more and more evidence is sought for the link between hypersympatheticotony and/or the lowered vagal tone and IHD progression.

The first studies in this respect indicate that the disturbed autonomic balance in the case of diagnosed IHD is associated with elevated risk of SCD. The decreased SCD is associated with poor prognosis after myocardial infarction. Farrel, Bashir, Cripps, et al. [1] stratify the risk of arrhythmia and sudden death in patients after myocardial infarction. They prove that the decreased SCD and the presence of late potentials are the most reliable predictors for the occurrence of dangerous ventricular arrhythmias and SCD. The risk of death after MI is 5.3 times higher in the group of patients with SDRR < 50 ms than in the group with SDRR > 50ms [2]. The decreased SCD also has an independent value for motivating the poor prognosis after infarction among the remaining risk-stratification tools: exercise ECG, ejection fraction, detection of ventricular ectopic activity and study of the late potentials [3].

The HRV time- and frequency-domain indices have the same value for predicting the risk of death after MI [4]. The authors have proven that the two sets of HRV indices evaluation are equivalent. According to them, VLF and LF are equivalent to SDRR, HF to RMSSD and PNN50. The indices from the time and frequency-domain are associated with risk of death during the observation of the patients after MI. According to [5], the frequency indices of total power, VLF significantly correlate, while LF and HF tend to correlate with the risk of all-cause mortality, cardiac death and death from arrhythmias after infarction. VLF has a prevalent predictive sensitivity in evaluation of the risk of rhythm death. HRV gives an opportunity to identify a small group of patients for whom the risk of death is 50% for 2.5 years. Bigger, et al. [6] are also the authors of a study of HRV in the early and late periods after MI. They conclude that the values of all frequency indices from HRV are significantly higher in norm compared to the values in acute and chronic IHD. The difference between healthy individuals and post-MI subjects is greater in the acute than in the chronic phase.

However, the studies in coronary disease without previous MI are not unequivocal. Rich, Saini, Kleiger, et al. [7] have studied the possibility the decreased HRV to serve as predictor of death also in patients with angiographically verified IHD, but without previous myocardial infarction. They have found that the decreased HRV and the low ejection fraction are the best predictors for the occurrence of death. Other studies report results specifying this conclusion. Wennerblom, Lurje, Tygesen, et al. [8] have studied patients with uncomplicated coronary artery disease. Compared to healthy individuals, the patients with angina pectoris demonstrate lower HRV values only for the indices of total power, HF, LF, RMSSD and PNN50. There is no significant difference in the values of the indices LF/HF, SDRR and SDARR. ***This result presupposes that the damage affects more the parasympathetic part of the autonomic balance.*** The absence of changes in SDRR and SDARR, which is associated with poor prognosis after MI, explains the good prognosis in this study in the patients with uncomplicated ischaemic heart disease. A probable reason for the different results is the different severity of IHD without previous MI in the groups of persons studied. In one group of patients the HRV indices indicate presence of hypersympatheticotony and there is also higher mortality among these persons. In another group of patients, the changes testify to impaired vagal tone. For these patients the prognosis is better. If HRV is evaluated separately for these two groups of persons, the patients with higher risk of death will be clustered more accurately compared to the results in the general inhomogeneous population.

The suppressed HRV is associated predominantly with risk of death as a result of rhythmic disorders. This assumption reflects the link between the disturbed autonomic balance and the electrical myocardial vulnerability. It is seen from the studies, however, that the damage in the autonomic balance has a much deeper connection with the pathophysiological changes occurring in the case of IHD. Tsuji, Larson, Venditti, et al. [9] have studied the link between the decreased HRV and the new cases of cardiac incidents in a large population from the Framingham Heart Study and Framingham Offspring Study. All patients in these studies lack clinical evidence for IHD or heart failure in the beginning of the study. The mean time of observation is 3.5 years. Researchers have found that SDRR, VLF, LF and total power significantly correlate with the growing risk of cardiac incidents. The actual cardiac incidents in the population for the period under review are ordered in frequency as follows: angina pectoris, recognized myocardial infarction, congestive heart failure, death of coronary heart disease, unrecognized myocardial infarction and coronary insufficiency. According to the results of the study, in most persons from the population with decreased HRV, the decrease was not a sign of rhythm death, but of complications of the coronary disease. Obviously, the reduced HRV as a factor predicting poor prognosis should not be associated only with the proarrhythmic effect of the hypersympatheticotony and/or with the decreased parasympathetic tone, which increase the risk of SCD. The lowered HRV is connected with all cardiac death. This conclusion is also confirmed by Singh, Mironov, Armstrong, et al. [10]. The authors have studied the prognostic power for poor outcome of the early measurements of HRV (in the first 48 hours) in the case of MI. They have also studied the influence of the different types of thrombolytic therapy, of the localization of the infarction, of the left-ventricular pump function, of the changes in the ST-segment and of the angiographically verified changes in the affected vessels on HRV. All indices of HRV from the time and frequency-domain, with the exception of the LF/HF ratio, decrease between the 1st and the 2nd day. There is no difference in the values of the HRV indices when the different thrombolytic strategies are applied. HRV is lower in anterior infarction, being higher for the less significant angiographic finds. An inverse correlation has been found to exist between the duration of the ST-changes and the frequency HRV indices. The low value of LF/HF on the first day after MI indicates the highest risk of death in the next 30 days to one year. These data again confirm the prognostic value of HRV after MI, but they also confirm the link between the decreased HRV (consequently, of the disturbed autonomic balance) and the ischaemia and atherosclerosis. Wennerblom, Lurje, Solem, et al. [11] have studied the possibility with revascularization using percutaneous transluminal coronary angioplasty (PTCA) to restore the disturbed HRV in patients with uncomplicated coronary artery disease. As we indicated earlier, these patients have lower values of the indices of total power, HF, LF, SDRR, RMSSD and PNN50 compared to the values in healthy individuals. There is no significant difference for the values of LF/HF, SDRR and SDARR. The complete revascularization is followed by a substantial recovery of the values of the indices reflecting the vagal activity – HF, RMSSD and PNN50 – but according to data from short recordings and with controlled respiration. No such recovery is observed with 24-hour recordings. The authors conclude that the disturbed HRV is not due to ischaemia only. The link between the disturbed autonomic balance and the ischaemia is confirmed also in [12]. According to the authors, the analysis of HRV, which was studied 30, 15, 5 and 1 minutes prior to the elevation of the ST-segment and during the peak of the elevation in patients with vasospastic angina pectoris, demonstrated the existence of a link between the disturbed autonomic balance and the spasm of the vessels in these patients. The influence of the suppressed parasympathetic tone in this group of patients is more pronounced. The values of HF decrease 2 minutes before the start of the ischaemia and are restored during the peak of the ischaemic incident. The heart rate and LF increase during the peak of the ischaemic incidents. Lanza, et al. [13] have studied the possibility to stratify better the risk of subsequent incidents in patients with unstable angina and have found that the frequency of the ischaemic episodes is four times higher in patients with value of the LH/FH ratio of 2.45 than in patients with LH/FH ratio of 1.3. As the LH/FH ratio is associated with the autonomic balance, it may be concluded that the hypersympatheticotony also involves higher probability for spasm of the vessels.

There exists a link between the disturbed autonomic balance and atherosclerosis. On the one hand, the hypersympatheticotony is associated with elevated levels of cortisol, catecholamines, serotonin, renin, angiotensin, aldosterone and the free radicals, with hyperlipidaemia and with impaired glucose utilization, and on the other – hypoparasympathicotonia is associated both with the low acetylcholinea level and with the deficit of nitrogen oxide, endorphins, coenzyme Q, antioxidants and other protective factors [14]. Thus, the combination of hypersympatheticotony with lowered vagal tone could contribute to an early start or could accelerate atherosclerosis. Huikuri, Jokinen, Syvanne, et al. [15] have studied the hypothesis that the decreased HRV is connected with the progression of atherosclerosis. The patients are with previous bypass-surgery with low HDL levels (< 1.1). One group was treated with placebo, the other one – with Gemfibrozil. The progression of atherosclerosis was monitored angiographically prior to and 32 months after the therapy. The patients on placebo with visible progression of atherosclerosis had the lowest SDRR values and the highest minimum heart rate, while those subjected to gemfibrozil therapy had no significant progression of atherosclerosis; in their case there was no link between the atherosclerotic lesions and HRV. Using multiple regression analysis, researchers have found that SDRR has a predictive value with respect to the development of focal atherosclerosis in the proximal parts of native coronary vessels. The decrease of SDRR there had no predictive value for the development of diffuse atherosclerotic changes and of new atherosclerotic lesions.

Obviously, the emergence and the progression of IHD are connected with the changes in the autonomic balance of the heart. Further studies in this field could improve the stratification of those patients with IHD for whom the risk of death is high. Impacting the damage in the autonomic balance with suitable drugs will improve the prognosis in such patients. If the effect of the disturbed autonomic balance on the appearance of IHD is clarified, it will be possible to have better prevention for the people with high risk of lethal outcome of this disease.

CIRCADIAN CHARACTERISTICS OF THE ISCHAEMIC INCIDENTS

It was indicated in *Chapter 4* that the frequency of cardiovascular incidents has a characteristic circadian nature with two peaks: a bigger peak in the morning and a second smaller one in the late afternoon hours.

Myocardial ischaemia in the presence of underlying coronary disease is caused by elevated oxygen demand of the myocardium and/or by disturbed oxygen supply. The oxygen supply is disturbed most frequently by reduction of the coronary blood flow as a result of the formation of a thrombus and/or vasospasm. In stable angina the leading cause is the elevated oxygen demand. In acute ischaemic incidents (unstable angina, MI) a more frequent cause is the impaired oxygen supply due to formation of a thrombus and/or due to vasospasm.

In stable angina, the higher oxygen demand in the morning hours are frequently due to higher heart rate and arterial blood pressure, as well as to the higher contractility of the myocardium, induced by the high level of adrenalin in the blood. Simultaneously with these endogenous factors, exogenous factors have also an influence – physical and mental activity rises sharply in the morning hours. This results in an unfavorable constellation of factors in the morning, increasing the oxygen demand of the myocardium and leading to ischaemic incidents. Krantz, Kop, Gabbay, et al. [16] have studied patients with stable angina with the aim of elucidating the link between the morning peak of myocardial ischaemia and the different triggering factors: physical and mental activity. These authors have found that identical in force exogenous stimuli cause more frequent and long ischaemic episodes in the morning hours compared to the other periods of the day and night. On the basis of this fact they conclude that myocardial ischaemia has both exogenous and endogenous circadian nature. The rising incidence of ischaemic incidents in the morning correlates with the elevated mean HR and with the peak HR in the hour preceding the onset of the incident. The authors note that the increased number of ischaemic episodes in the morning persists even after correction of the mean HR in the hour preceding the incident, as well as after correction of HR during the ischaemic episode. Deedwania and Nelson [17] have found elevation of the arterial blood pressure in the morning hours before the onset of the ischaemic incidents in patients with stable coronary disease. Parker, Testa, Jimenez, et al. [18] have studied the circadian characteristic of the ischaemic episodes in the case of stable angina pectoris and their dependence on the physical activity. The studies were conducted in two successive days. On the first day the patients had normal physical activity, waking and moving at 8 a.m. On the second day they woke up at 8 a.m. but got up at 10 a.m. A part of the patients were treated with the beta-blocker Nadolol, the rest were on placebo. The patients on placebo were found to have a morning increase of the frequency of ischaemic incidents, corresponding to the start of the physical activity on the first day. On the day with delayed physical activity there was a 4-hour shifting of the peak of the ischaemic incidents, which proves their link with the exogenous cause. In 87% of the patients on placebo there was HR elevation before the ischaemic episodes. In the patients treated with Nadolol there was a 50% reduction of the ischaemic episodes and they had no morning peak. Although the therapy with beta-blocker leads to significant reduction of the total number of ischaemic episodes, there was also a significant increase of the ischaemic episodes, which were not preceded by elevated HR. The probable reason for this is the vasoconstrictor effect of Nadolol.

The link between the circadian nature of the ischaemic patients and that of the autonomic balance in stable angina is studied in [19]. The authors prove that in the case of stable angina the ischaemic episodes and HRV have a similar circadian characteristic. The peak of the ischaemic incidents and of the LF/HF ratio is during the daytime hours, whereas the values of the indices LF and HF are maximal during the night when ischaemic incidents are the fewest. Wennerblom, Lurje, Karlsson, et al. [20] have also studied the circadian nature of the autonomic balance in patients with angina pectoris, divided in three groups. The first group is without therapy, the second – on therapy with nitrates, the third – on therapy with beta-blockers. The results are compared with the data in healthy controls. The authors find that there is circadian nature in the vagal tone in healthy individuals, determined by studying HF. This circadian nature strongly decreases in angina pectoris patients. The treatment with nitrates and beta-blockers tends to normalize the circadian characteristic of the vagal tone and simultaneously with this it reduces the hypersympatheticotony in the morning hours in angina pectoris patients. Other authors have found that the circadian nature of the autonomic balance in patients with stable angina is reduced. Huikuri, Niemela, Ojala, et al. [21] have studied the circadian characteristic of HRV in patients with stable angina compared to healthy individuals. The latter have a significant circadian rhythm, determined through the normalized values of HF. In them the HF values are the highest during sleep. The normalized values of LF and the LF/HF ratio also have a circadian rhythm in healthy individuals, with the highest values during the day. At the same time, no circadian periodicity is found in the values of all spectral indices for HRV in patients with stable angina. In the morning hours, standing up causes a significant rise in the LH/FH ratio in healthy individuals. In persons with stable angina there was no change in the values of the ratio after standing up. According to the authors, this result is explained with impaired capacity of the autonomic nervous system to respond to stimuli in patients with IHD.

As was mentioned already, reduced oxygen supply due to formation of a thrombus and/or vasospasm is a major pathophysiological cause for the acute ischaemic incidents (unstable angina (UA), MI). As the vasospastic hormones, the autonomic balance, coagulation and fibrinolysis have circadian characteristics (see *Chapter 4*), it would be logical to detect circadian nature also in acute coronary syndromes. Figueras and Lidon [22] have studied patients with unstable angina pectoris. Using intraatrial stimulation, they have found that unstable angina has a characteristic circadian model of the ischaemic episodes, with a peak in the morning hours. The same peak is also noted in patients who remain lying in the morning hours due to the high severity of their disease. BP and HR are higher in the noon hours compared to the morning. This result suggests that the elevation of HR and BP, as well as the exogenous activity, are not decisive for the appearance of circadian nature. Probably the elevated cardiovascular tone is a triggering factor. The circadian characteristic in the frequency of the ischaemic incidents in unstable angina is also confirmed in [23]. The authors have studied patients included in the registers of TIMI III and TIMI IIIB (Thrombolysis in Myocardial Ischemia) and have found that the frequency of the incidents is the highest between 6 a.m. and noon.

Circadian characteristic of the ischaemic episodes has also been detected in vasospastic angina (variant angina). The morning peak in the frequency of the incidents is slightly earlier than the one observed in the other IHD types. Lanza, Patti, Pasceri, et al. [24] have studied 26 patients with variant angina by 24-hour ECG monitoring. Circadian variation of the ischaemic incidents is found in 14 patients without significant stenoses. The peak of the incidents was around 2:30 a.m. There was no circadian nature in the occurrence of the ischaemic episodes in the patients with significant stenoses.

In spite of the obvious link between the circadian characteristic of the incidents in IHD and the autonomic balance, we failed to find data in the literature on the circadian nature of HRV in unstable coronary disease.

The unfavorable combination of factors that can induce coronary vasospasm in the morning hours exists both in healthy individuals and in patients with IHD. Obviously, in healthy individuals there is protection against this unfavorable constellation. Such a protective factor is the intact vascular endothelium. The coronary endothelium plays a key role in the regulation of the coronary tone, and platelets adhesion and aggregation. It is important to note that the normal myocardial vessels and the vessels altered by atherosclerosis or even only with endothelial dysfunction have a different response to vasoactivating stimuli. Hassan El-Tamimi, Mansour, Pepin, et al. [25] have studied the dilatory response of the coronary vessels with normal and with dysfunctional endothelium, as well as the circadian characteristic of the tone of the myocardial vessels in patients with cardiac artery diseases (CAD). The segments with normal endothelium respond with dilation to acetylcholine infusion, whereas the segments with endothelial dysfunction respond with constriction. In the segments responding to acetylcholine with vasoconstriction, there exists a significant difference between the degree of the constrictor response in the morning hours (when it is maximal) and the response during other intervals of the day and night. In the group with normal endothelium there is no significant difference in the dilatory response to infusion of acetylcholine in dependence on the time of the day. This proves that the vasomotor response to stimulation depends on the normal function of the endothelium. Probably other stimuli, too, can induce vasoconstriction in the case of endothelial dysfunction, while in a normal vessel the response would be vasodilation.

Uren, Crake, Tousoulis, et al. [26] have also found a changed response to stimulation of the vessels in the case of IHD. They have studied the response to cold pressor stress of the collateral-dependent myocardium. It is known that cold stimulation is a typically sympathetic test. In normal vessels it causes a decrease and in collaterals – an increase of vascular resistance. This study proves that the altered vessels have a pathological response to sympathetic stimulation. Shaw, Chin-Dusting, Kingwell and Dart [27] have also studied the response of vessels to stimulation in healthy individuals and in patients with IHD, as well as the probable circadian nature in the reactivity of the vascular response to vasoactive stimuli. Simultaneously with this, they have measured other indices as well: HR, BP, spectral characteristics of HRV, plasma cortisol and markers of the inflammation. The healthy individuals have the highest value of HR and LF from HRV in the morning hours, which presupposes a peak of sympathetic activity. A specific circadian characteristic of the basal forearm blood flow has been discovered in the controls, with significant reduction of the blood flow at 8 p.m., compared to 8 a.m. and 2 p.m. In the case of acetylcholine infusion, the dilatory response was stronger at 8 a.m. compared to 8 p.m. The patients with CAD showed no difference in the basal forearm blood flows in the different hours before and after the application of acetylcholine. In healthy individuals, the diurnal characteristic of the endothelium-dependent dilation presupposes that the normal endothelium is capable of resisting the potential vasoconstrictor factors, which have an unfavorable constellation in the morning hours.

The cited studies indicate that in the case of IHD the vasoactive response is impaired and this causes under certain unfavorable circumstances vasoconstriction and provokes the appearance of ischaemic incidents. The normally functioning endothelium guarantees normal response to vasoactivating stimuli and is part of the protective mechanism of the healthy heart against the risk factors in the morning hours.

Another unfavorable risk factor influencing the circadian nature of the pathophysiology of the ischaemic incidents is the procoagulant activity found in the morning hours. Already back in the 1980s, Tofler, Brezinski, Shafer, et al. [28] found that the platelets aggregation, when the aggregation reagents adenosine and epinephrine are applied, is the highest between 6 and 9 a.m. Brezinski, Tofler, Muller, et al. [29] add that the accelerated aggregation in the morning hours with respect to adenosine and epinephrine is observed only when the patients stand up. The blood viscosity also increases in the morning [30]. In addition to this unfavorable tendency, white blood cells also tend to aggregate more in the morning [31]. The procoagulant tendency is also combined with reduced fibrinolytic activity in the morning. Masuda, Ogava, Miyao, et al. [32] have studied the circadian characteristic of the fibrinolytic activity in patients with variant angina, with stable angina and in healthy individuals. Key components of fibrinolysis are the tissue plasminogen activator, which promotes fibrinolysis, and its specific inhibitor the plasminogen activator inhibitor. The authors have found that the PAI and t-PA-antigen concentrations are the highest in the morning hours in the three groups of patients, but in the group with variant angina the concentrations are the highest. This result suggests that fibrinolysis not only decreases in the morning, but also that in acute coronary syndromes this unfavorable tendency is aggravated. The importance of the platelet function is also emphasized by the Physicians Health Study [33], where the morning peak of myocardial infarctions is reported to be lower for the patients subjected to aspirin therapy.

It was indicated that the elevation of HR and blood pressure (BP) has the highest importance for the morning peak in the frequency of the coronary incidents. In addition to leading to a rise in the oxygen consumption of the myocardium, BP also increases the shear forces in the vessel and may cause a fissure or rupture of the plaque. The elevated coronary vascular tone due to hormonal stimuli with peak in the morning hours also leads to increased risk of vascular incidents in the morning, as well as procoagulant activity of the blood and decreased fibrinolysis. These factors potentiate the circadian nature in myocardial infarction as well. Muller, Stone, Zoltan, et al. [34] have studied the occurrence of MI during the day and during the night according to data from MILIS and have found significant circadian nature of the myocardial incidents with a peak in the interval from 6 a.m. until noon. By studying cardiac muscle isoenzyme of creatine kinase (CK-MB) they have found that around 9 a.m. there is a threefold increase of the incidents. There is no circadian nature in the occurrence of MI in the patients treated with beta-blocker. Willich, Linderer, Wegscheider, et al. [35] have found the time of occurrence of MI in patients covered by ISAM. The MI incidents increase considerably in the interval between 8 a.m. and noon, and occur 3.8 times more frequently between 8 and 9 a.m., compared to between midnight and 1 a.m. Here, too, the patients treated with beta-blocker did not have a peak of the incidents in the morning hours. The already cited study [33] (Physicians Health Study) covers 22,071 persons, part of whom on therapy with aspirin, the rest – on placebo. They have found that the group on placebo has bimodal distribution of the frequency of the MI incidents in the day and night, with first peak in the morning between 4 and 10 a.m. and a second smaller peak in the evening hours. Myocardial infarctions are fewer in the group treated with aspirin and there is no peak in the morning hours. Obviously, with suitable drug therapy it is possible to influence the factors determining the circadian nature, and to reduce the risk of ischaemic incidents in the morning hours. Goldberg, Bradi, Muller, et al. [36] have studied the time of occurrence of MI related to the morning activity and have found that approximately 23% of the patients report onset of the symptoms approximately one hour after awakening. This link of the sudden physical activity with the beginning of the ischaemic incidents stems from the changed reactivity of the autonomic nervous system to stress in patients with cardiovascular diseases. On the other hand, the changes in the autonomic balance influence the heart rate, the arterial blood pressure and the reactivity of the vessels.

The cited circadian model of the ischaemic incidents is impaired in some more special cases of MI.

For example, in patients with non-Q wave MI there exists a nocturnal peak of the incidents. An analogous circadian model is also observed for patients with previous heart failure, peripheral vascular disease or stroke [37]. Rana, Mukamal, Morgan, et al. [38] note a change in the circadian characteristic of the MI incidents. They have studied the time of occurrence of the myocardial infarction in patients without diabetes, in patients with type I diabetes and in patients with type II diabetes, with different duration of the disease. In the patients without diabetes and these with type II diabetes with duration less than five years the circadian characteristic is similar, with a peak of MI in the morning hours. The patients with type I diabetes and these with duration of the type II diabetes exceeding five years had no circadian nature in the distribution of the frequency of infarctions. One of the explanations that the authors give for this absence of circadian nature of MI in the case of prolonged diabetes is the presence of autonomic neuropathy. Diabetic autonomic neuropathy is known to damage first the parasympathetic part of the autonomic control. There is evidence of reduced activity of the parasympathicus during sleep in diabetes patients. Probably the impaired circadian characteristic of the autonomic balance are at the basis of the missing circadian nature in the case of diabetes.

The link between the circadian nature of the autonomic balance and the changes occurring in the case of MI are very poorly investigated. Malik, Farrell and Camm [39] have studied the circadian characteristic of HRV after MI in two groups of patients. The first group consists of persons with serious complications in the postinfarction period, the second – persons without complications for a period of more than six months after the incident. They have found that in the low-risk group there are more pronounced diurnal changes in the HRV, i.e., the circadian nature is almost preserved in it, unlike HRV in the high-risk group.

From the survey made it may be concluded that *the circadian characteristic of the autonomic balance in the acute coronary syndromes is almost unstudied.* Research in this field could contribute to clarifying the role of the autonomic balance in the circadian nature of ischaemic incidents. It is possible to differentiate specific groups of patients according to the damage to the autonomic balance. This would improve, on the one hand, the risk stratification of the patients with IHD, and on the other – it would help modify the drug therapy for the changes in the circadian nature of the autonomic balance. Such a therapy would reduce the risk of cardiovascular incidents and would improve the long-term prognosis in patients with IHD.

CIRCADIAN NATURE OF THE AUTONOMIC BALANCE IN UNSTABLE ANGINA PECTORIS. PROTECTIVE ROLE OF THE VAGAL DOMINATING ACTIVITY

We studied 26 patients with unstable angina pectoris between 6 and 10 days after the beginning of the complaints. The diagnosis unstable angina was determined according to the classification of the Canadian Cardiovascular Society (CCS). The patients were studied during their hospitalization. Nine of the patients reported preceding hypertension. The therapy applied to the patients differed. It included nitrate drugs, calcium antagonists, ACE-inhibitors and antithrombotic agents. The patients subjected to preceding therapy with beta-blocker had at least 48 hours without administration of the blocker before the study of HRV.

The exclusion criteria were diabetes, heart failure, previous myocardial infarction, renal failure and other general severe diseases.

All patients are in sinus rhythm without conduction disorders.

The values of the HRV indices during rest, with the handgrip test and with the Valsalva manoeuvre in the morning and in the afternoon in the patients with UA are significantly lower ($p<0.05$) than the values in healthy individuals – see Tables 7.1, 7.2 and 7.3 and Tables 6.2, 6.3 and 6.5. This is evidence of impaired autonomic balance in this group of patients.

The comparison between the values of the HRV indices, measured in the morning and in the afternoon in the patients with UA, did not reveal a significant difference in all three samples: during rest, with the handgrip test and with the Valsalva manoeuvre (Tables 7.1, 7.2 and 7.3). The result is analogous to the result in healthy individuals.

Table 7.1. Mean values of the HRV indices in the morning (M) and in the afternoon (AN) during rest in patients with UA.

	Mean(M)	SD(M)	Mean(AN)	SD(AN)	t value	p<
SDRR	24.15	7.4481	25.62	7.6217	0.4945	n.s.
MDRR	18.54	5.6364	19.54	6.1727	0.4313	n.s.
dSDRR	8.69	3.8597	9.15	3.6934	0.3115	n.s.
dMDRR	6.54	3.1521	7.00	2.7689	0.3966	n.s.
PNN50	0.62	1.3868	0.62	0.6504	0.0000	n.s.
RMSSD	13.38	6.3316	14.85	5.8998	0.6089	n.s.
VLF	76.31	60.8062	104.77	105.3361	0.8437	n.s.
LF	38.77	28.5341	54.54	38.7742	1.1810	n.s.
HF	28.15	27.9280	32.62	32.9887	0.3722	n.s.
LF/HF	2.01	1.2289	2.80	1.8773	1.2744	n.s.

In the group of patients with UA with the predominantly sympathetic sample – handgrip test – we found relations in the values of the HRV indices compared to their values during rest that we have never registered in healthy individuals and in patients with other CVD. This is valid of the comparisons between the data from the studies both in the morning and in the afternoon (Tables 7.1 and 7.2). For dSDRR, dMDRR, PNN50, RMSSD and LF/HF there was no difference between the values during rest and with the handgrip test. For SDRR and MDRR with the test we observed an unusual change in the values: they were higher than the values during rest. In healthy individuals and in the groups of patients with other CVD, the values of these indices with the handgrip test always decreased – a normal response because the handgrip test stimulates the sympathicus, which leads to reduction of HRV. In this sense, the response to the stress test in patients with HCA is "pathological". Its explanation should be sought in the complex possible reactions of the two parts of ANS, including also lack of response by the sympathetic part to its incapacity to react, i.e., blocking of the sympathicus during stress with hypersympatheticotony. In other words, the sympathetic autonomic dysfunction is of a type that allows domination of the parasympathicus during stimulation. The tendency the vagus to manifest protective prevalence during sympathetic stimulation explains the extremely good prognosis in the studied population of patients. Among all patients with UA there was only one lethal outcome and there was no aggravation of IHD during the three-year follow-up.

Comparison between the values of the frequency indices VLF, LF and HF with the handgrip test and rest showed that the results in the patients with UA were similar to those in healthy individuals (see Tables 6.2 and 6.3). In the morning the response to stimulation was weaker than in the afternoon due to the higher sympathetic tone in the morning, limiting the possibility of additional stimulation, and the lower sympathetic tone in the afternoon, allowing a higher level during stimulation.

The mean value of TRABI for the HRV indices from the comparative study of the response to the handgrip test with respect to rest in the morning and in the afternoon in the patients with UA is higher, although insignificantly, than the respective value in healthy individuals. For certain indices the values of the index in the patients considerably exceeded the values in the healthy individuals (Table 7.4). Such are the assessments for SDRR, dSDRR, dMDRR, PNN50 and the LF/HF ratio (Figure 7.1). *These results indicate that in patients with UA the sympathetic component in AB is more time-dependent than in healthy individuals.*

Table 7.2. Mean values of the HRV indices in the morning (M) and in the afternoon (AN) with the handgrip test in patients with UA.

	Mean(M)	SD(M)	Mean(AN)	SD(AN)	t value	p<
SDRR	27.31	10.9953	28.69	15.5209	0.26246	n.s.
MDRR	21.85	9.6768	22.31	12.2228	0.10674	n.s.
dSDRR	8.15	4.1802	9.15	4.0383	0.62034	n.s.
dMDRR	6.08	3.2265	6.92	3.2006	0.67131	n.s.
PNN50	0.54	1.1983	0.77	1.1658	0.49770	n.s.
RMSSD	13.08	7.0529	14.62	6.5516	0.57623	n.s.
VLF	38.23	35.7192	40.46	45.9757	0.13815	n.s.
LF	8.08	4.1726	13.08	8.2408	1.95171	n.s.
HF	6.54	6.9476	9.08	8.3612	0.84192	n.s.
LF/HF	2.35	2.0575	2.32	1.2285	0.04977	n.s.

Table 7.3. Mean values of the HRV indices in the morning (M) and in the afternoon (AN) with the Valsalva manoeuvre in patients with UA.

	Mean(M)	SD(M)	Mean(AN)	SD(AN)	t value	p<
SDRR	59.69	32.8288	59.46	30.2176	0.0186	n.s.
MDRR	47.69	28.0695	43.46	23.5783	0.4161	n.s.
dSDRR	20.38	10.8362	20.92	9.8358	0.1327	n.s.
dMDRR	12.92	5.7802	13.77	6.2203	0.3593	n.s.
PNN50	3.92	3.0676	4.23	4.3618	0.2080	n.s.
RMSSD	26.46	12.1355	28.08	11.6438	0.3463	n.s.
LF	131.62	192.8292	145.46	268.4401	0.1510	n.s.
HF	19.38	35.7690	26.77	47.9186	0.4453	n.s.
LF/HF	13.57	10.7482	6.98	5.1352	1.9975	n.s.

Table 7.4. Values of TRABI for the HRV indices in patients with UA and in healthy individuals.

HRV indices	Unstable	Angina	Healthy	Subjects
	RS vs. HG	RS vs. VM	RS vs. HG	RS vs. VM
SDRR	0.300	0.200	0.045	0.182
MDRR	0.000	0.200	0.068	0.182
dSDRR	0.300	0.100	0.091	0.159
dMDRR	0.300	0.100	0.114	0.091
PNN50	0.300	0.000	0.159	0.159
RMSSD	0.100	0.100	0.159	0.227
VLF	0.000		0.091	
LF	0.100	0.400	0.045	0.045
HF	0.000	0.200	0.000	0.091
LF/HF	0.400	0.600	0.227	0.136

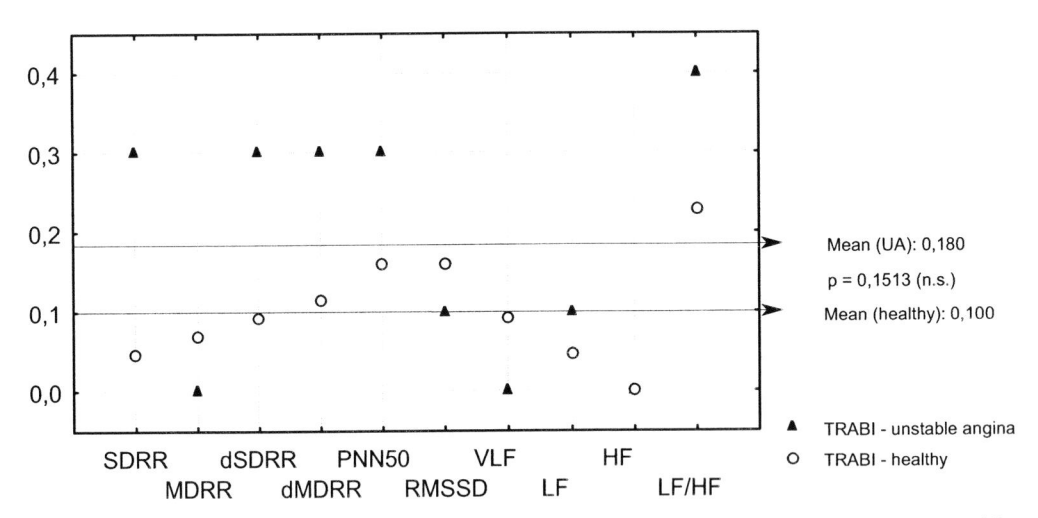

Figure 7.1. Values of TRABI for the HRV indices in healthy individuals and in patients with UA with comparison between morning and afternoon measurements during rest and with the handgrip test.

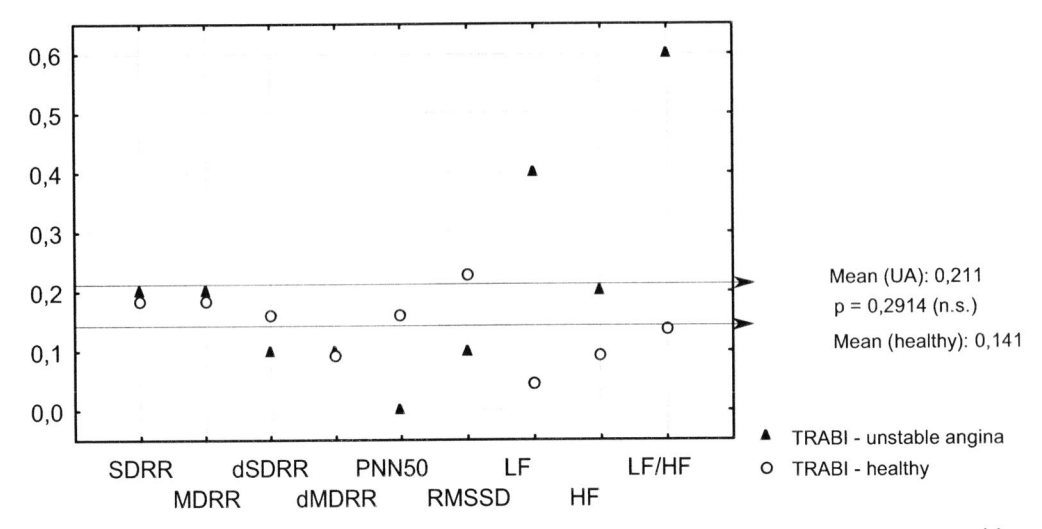

Figure 7.2. Values of TRABI for the HRV indices in healthy individuals and in patients with UA with comparison between morning and afternoon measurements during rest and with the Valsalva manoeuvre.

In the group of patients with UA, the comparison between the values of the HRV indices during rest and with the Valsalva manoeuvre in the morning demonstrates considerable response compared to the parasympathetic sample. In the morning the vagal tone is low and permits stimulation. The response in the afternoon hours is weaker, because the parasympathicus has attained a relative stability. There is a pronounced circadian nature of the parasympathetic part in the autonomic balance in the population of patients with UA.

The mean value of TRABI for the HRV indices from the comparative study of the response to the Valsalva manoeuvre and rest in the morning and in the afternoon in the patients with unstable angina does not differ significantly from the value in healthy individuals (Figure 7.2). For most indices from the time area the values of TRABI in patients with UA and in healthy individuals are close (Table 7.4). ***This result is in support of the conclusion that the circadian characteristic of the parasympathetic tone in persons with UA is analogous to the normal one.*** However, the values of the index for the indices LF and LF/HF are higher in patients with UA than in healthy individuals, being 0.4 for LF and reaching 0.6 for LF/HF. Probably in the sample in which there is a sympathetic component as well, the stimulation of the sympathicus is suppressed due to its blocking described above. As a result, the frequency indices respond in a range in which there was no saturation of the values, and the time changes in the autonomic control can be reflected.

The results from the study in patients with UA indicate the existence of autonomic dysfunction with blocked sympathetic activity. The evaluations of the HRV indices with TRABI show a slightly more pronounced circadian nature of sympatheticotony than in healthy individuals. However, the parasympatheticotony in the patients with UA has a considerably pronounced circadian nature, similar to the circadian nature in healthy controls. During stress (blocking the sympathicus), the parasympathicus performs a protective role and thus secures a favorable long-term prognosis for the patients with UA in the population studied. It can be assumed that the protecting parasympathetic activity was missing in patients with unstable angina pectoris who suffered from aggravation of the IHD or a lethal outcome occurred.

RELATION BETWEEN THE CIRCADIAN NATURE OF THE AUTONOMIC BALANCE AND THE CIRCADIAN NATURE IN THE INCIDENTS FROM MYOCARDIAL INFARCTION

We divided the patients with myocardial infarction into two groups to determine the link between the circadian nature of the autonomic balance and the circadian nature of the cardiovascular incidents (CVI). The first group comprises the patients for whom MI occurred in the morning between 6 and 11 a.m, the second – the patients for whom the incident occurred at a different time. The group with morning infarction had 15 patients. The group with MI at a time different from the morning consisted of 19 patients. There was no difference in the distributions of the localizations of the infarction in the two groups. The exclusion criteria comprised patients with diabetes, frequent rhythmic disorders, conduction disorders, heart failure and other severe general diseases. The patients were subjected to outpatient follow-up examinations between the 3rd and the 5th month after the incident. There was no difference in the type of drugs administered to both groups of patients. The therapy with beta-blockers was discontinued at least 7 days prior to the study. No patient was on anriarrhythmic therapy.

Our working hypothesis was that the patients in the two groups would have different circadian characteristics of AB. The confirmation of this hypothesis will contribute to clarifying the role of AB for the occurrence of cardiovascular incidents (CVI).

Results in the Patients with Morning Myocardial Infarction (MMI)

The mean value of TRABI for the HRV indices from the comparative study in the morning and in the afternoon of the response to the handgrip test compared to rest for the patients with MMI is significantly higher than for healthy individuals (Figure 7.3). *It can be claimed that the patients with MMI have a marked circadian characteristic of the sympathicus.* For all indices, with the exception of LF/HF, the values of TRABI are higher in the patients with MI. The lower value of the index for LF/HF is explained with hypersympatheticotony and blocked response to stress, on the one hand, and decreased circadian nature of the parasympathicus – on the other, which does not allow this indice to reflect the time changes in AB.

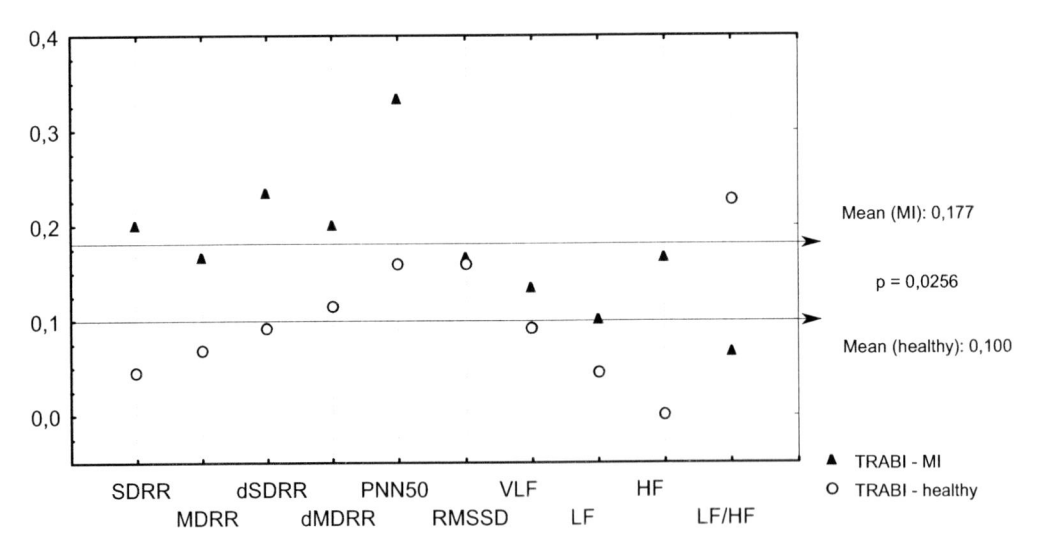

Figure 7.3. Values of TRABI for the HRV indices in healthy individuals and in patients with MMI with comparison between morning and afternoon measurements during rest and with the handgrip test.

The mean values of the actual HRV indices in the patients with MMI during rest, with the handgrip test and with the Valsalva manoeuvre are significantly lower than the mean values in healthy individuals – see Tables 7.5, 7.6 and 7.7, and Tables 6.2, 6.3 and 6.5. The comparison between the values of the indices in the morning and in the afternoon during rest, with the handgrip test and with the Valsalva manoeuvre in the patients with MMI did not demonstrate significant differences. The comparison between the values of the indices in the morning and in the afternoon during rest and with the handgrip test indicates that the response to stress was low and almost identical in both time intervals. We have described analogous autonomic dysfunction of the sympathicus also in patients with unstable angina. Hypersympatheticotony can be assumed in the patients with MMI, which does not allow to discover a difference in the reactivity of the sympathicus in the different time intervals, hence the circadian nature is detected only by means of TRABI.

The mean value of TRABI for the HRV indices from the comparative study in the morning and in the afternoon of the response to Valsalva manoeuvre compared to rest in the patients with MMI does not differ significantly from the value in healthy individuals (Figure 7.4). The patients preserve the normal circadian nature of the parasympathetic component in

AB. The evaluations of TRABI for the indices SDRR, RMSSD and LF/HF are lower in patients with MMI. The result can be explained with the relatively decreased circadian nature of the parasympathetic part of the autonomic balance compared to the marked circadian nature of the sympathetic part in this group of patients.

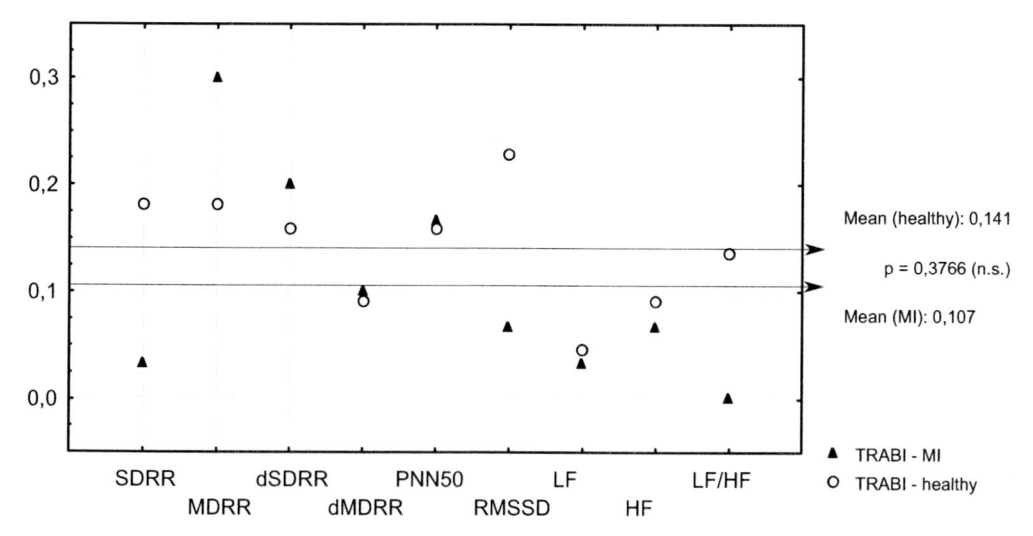

Figure 7.4. Values of TRABI for the HRV indices in healthy individuals and in patients with MMI with comparison between morning and afternoon measurements during rest and with the Valsalva manoeuvre.

The response of the parasympathicus in the afternoon hours is stronger than the response in the morning hours. This is evident from the comparison between the mean values of the HRV indices in the morning during rest and with the Valsalva manoeuvre and in the afternoon during rest and with the Valsalva manoeuvre in the patients with MMI. Earlier studies linked the better reactivity of the vagus with its lower tone. However, the manifested capacity for a higher response to stimulation in the afternoon hours in this case should probably not be associated with lower vagal tone, but rather with lower sympathetic tone, which allows a stronger response to stimulation.

Table 7.5. Mean values of the HRV indices in the morning (M) and in the afternoon (AN) during rest in patients with MMI.

	Mean(M)	SD(M)	Mean(AN)	SD(AN)	t value	$p<$
SDRR	24.87	13.9533	24.73	12.2327	0.0278	n.s.
MDRR	19.13	11.0316	18.27	8.8031	0.2378	n.s.
dSDRR	18.73	28.6767	11.20	7.6176	0.9833	n.s.
dMDRR	8.73	5.9817	7.67	4.8648	0.5358	n.s.
PNN50	2.60	3.8508	1.67	2.8452	0.7550	n.s.
RMSSD	18.67	13.4837	16.87	10.3155	0.4106	n.s.
VLF	98.27	98.4280	96.33	77.4409	0.0598	n.s.
LF	88.93	139.9298	66.00	70.6561	0.5666	n.s.
HF	60.13	76.2841	47.60	52.1561	0.5253	n.s.
LF/HF	2.64	3.3807	2.40	2.6944	0.2090	n.s.

Table 7.6. Mean values of the HRV indices in the morning (M) and in the afternoon (AN) with the handgrip test in patients with MMI.

	Mean(M)	SD(M)	Mean(AN)	SD	t value	p<
SDRR	20.33	8.6079	21.73	11.4109	0.3793	n.s.
MDRR	16.00	6.5465	15.87	8.4165	0.0484	n.s.
dSDRR	8.93	4.5429	19.13	29.3595	1.3297	n.s.
dMDRR	6.87	3.5630	8.00	5.2372	0.6930	n.s.
PNN50	0.67	1.5430	2.00	3.1623	1.4676	n.s.
RMSSD	14.40	7.5857	18.07	13.0902	0.9386	n.s.
VLF	26.87	35.1199	25.60	25.4945	0.1130	n.s.
LF	15.80	14.3238	24.67	40.3302	0.8024	n.s.
HF	12.13	10.7429	12.33	14.9316	0.0421	n.s.
LF/HF	3.24	4.8764	5.53	9.2812	0.7617	n.s.

Table 7.7. Mean values of the HRV indices in the morning (M) and in the afternoon (AN) with the Valsalva manoeuvre in patients with MMI.

	Mean(M)	SD(M)	Mean(AN)	SD(AN)	t value	p<
SDRR	50.73	30.9457	53.60	33.7084	0.2426	n.s.
MDRR	39.33	22.5505	43.00	26.1998	0.4108	n.s.
dSDRR	17.40	13.4366	15.87	11.4134	0.3368	n.s.
dMDRR	12.13	8.7003	11.47	8.4757	0.2126	n.s.
PNN50	3.73	4.6517	4.60	6.5115	0.4194	n.s.
RMSSD	23.47	16.4137	22.33	15.5456	0.1942	n.s.
LF	84.60	56.4912	117.67	144.2090	0.8269	n.s.
HF	19.87	34.3529	25.40	44.8247	0.3795	n.s.
LF/HF	29.14	34.2551	21.78	24.4577	0.6128	n.s.

The mean value of TRABI in healthy individuals is 0.100 in the comparative study between HRV during rest and with the handgrip test, and 0.141 in the study during rest and Valsalva manoeuvre. The result shows that the circadian characteristic of the parasympathicus in healthy individuals is more pronounced. The results in the patients with MMI are very different. In the first study, TRABI had a mean value of 0.177, in the second – 0.107. Consequently, abnormal circadian nature of the autonomic balance is observed in the patients with MMI. The circadian nature in these patients is more pronounced for the sympathetic part of the balance, unlike the circadian nature in healthy individuals.

Both in this group of patients and in the patients with unstable angina we found sympathetic autonomic dysfunction with hypersympatheticotony and blocked response to stress. Unlike the unstable angina, here the preventive function of the vagus is lacking. *The abnormal circadian characteristic of the autonomic balance and the sympathetic autonomic dysfunction during stress are probably the reason for the peak in the frequency of cardiovascular incidents in the morning.*

Results in the Patients with Myocardial Infarction in the Non-morning Hours (NMMI)

The mean value of TRABI for the HRV indices from the comparative study in the morning and in the afternoon of the response to handgrip test compared to rest in the patients with NMMI almost coincided with the value in healthy individuals (Figure 7.5). This result means that *in the patients with NMMI there was no substantial impairment in the circadian nature of the sympathetic part of the balance.* Only for the indices SDRR and MDRR the values of the index are higher in patients with NMMI compared to the evaluations in healthy individuals. Obviously, the circadian nature of the parasympathicus is impaired and hence also partial manifestation of the circadian nature of the sympathetic part of the balance is possible.

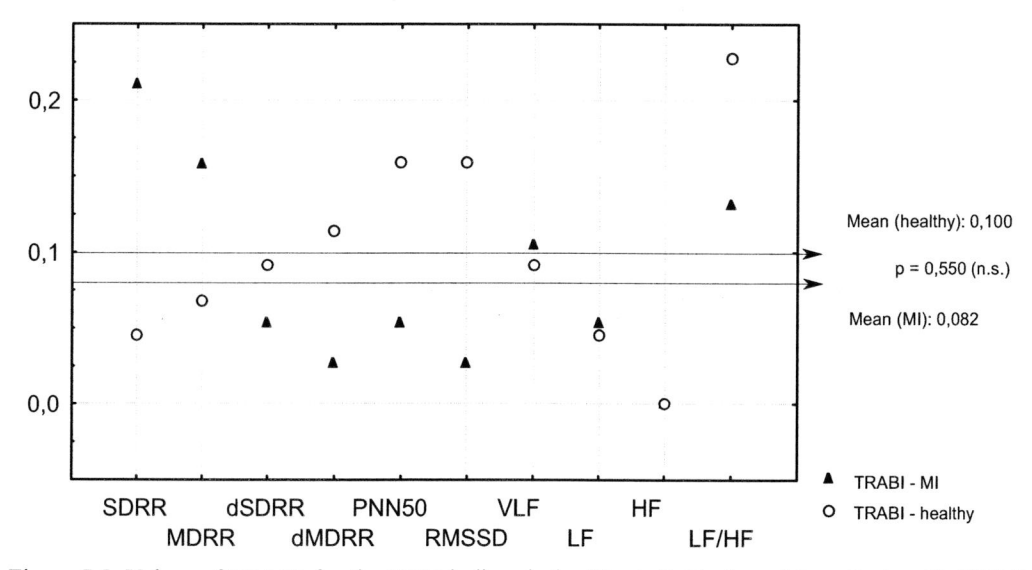

Figure 7.5. Values of TRABI for the HRV indices in healthy individuals and in patients with NMMI with comparison between morning and afternoon measurements during rest and with the handgrip test.

As with the morning infarctions, the values of the HRV indices during rest are significantly lower in patients with NMMI compared to healthy individuals – see Tables 7.8 and 6.2. Obviously, the autonomic control is impaired in the patients from this group as well.

There is no significant difference between the values of the indices in the morning and in the afternoon during rest, with the handgrip test and with the Valsalva manoeuvre in the patients with NMMI (Tables 7.8, 7.9 and 7.10).

Comparison between the values of the HRV indices during rest and with the handgrip test in the morning and accordingly in the afternoon in the patients with NMMI reveals the following specificities in the autonomic balance. A weaker capacity for response to the handgrip test is observed in the morning hours as a result of the higher sympathetic tone, whereas in the afternoon hours the response capacity is a little more pronounced. This reactivity of the sympathicus resembles the reactivity in healthy individuals.

Table 7.8. Mean values of the HRV indices in the morning (M) and in the afternoon (AN) during rest in patients with NMMI.

	Mean(M)	SD(M)	Mean(AN)	SD(AN)	t value	p<
SDRR	28.53	9.8959	27.60	15.3652	0.2209	n.s.
MDRR	20.89	6.8142	21.36	10.9482	0.1566	n.s.
dSDRR	19.11	25.7162	14.16	20.6889	0.6534	n.s.
dMDRR	9.11	4.5570	14.16	20.6889	1.0396	n.s.
PNN50	1.47	1.7438	2.63	5.4489	0.8822	n.s.
RMSSD	20.32	11.2746	20.74	23.3210	0.0709	n.s.
VLF	100.21	58.5478	120.84	162.0824	0.5218	n.s.
LF	138.79	322.4086	70.89	86.3403	0.8867	n.s.
HF	63.89	65.3162	84.84	173.1359	0.4934	n.s.
LF/HF	23.69	41.4611	19.27	36.8971	0.3476	n.s.

Table 7.9. Mean values of the HRV indices in the morning (M) and in the afternoon (AN) with the handgrip test in patients with NMMI.

	Mean(M)	SD(M)	Mean(AN)	SD(AN)	t value	p<
SDRR	21.32	8.8195	19.53	7.4040	0.6774	n.s.
MDRR	16.79	7.1770	15.32	5.9260	0.6902	n.s.
dSDRR	11.37	9.4470	14.05	25.7282	0.4269	n.s.
dMDRR	8.42	6.9227	7.11	5.0210	0.6707	n.s.
PNN50	1.95	5.9111	1.42	3.4851	0.3343	n.s.
RMSSD	17.84	13.3219	14.26	9.6312	0.9490	n.s.
VLF	25.00	16.7199	21.79	20.3349	0.5316	n.s.
LF	18.84	16.1701	18.11	23.3521	0.1131	n.s.
HF	12.79	14.6802	8.32	10.6981	1.0735	n.s.
LF/HF	7.82	22.6777	15.01	30.9852	0.8159	n.s.

Table 7.10. Mean values of the HRV indices in the morning (M) and in the afternoon (AN) with the Valsalva manoeuvre in patients with NMMI.

	Mean(M)	SD(M)	Mean(AN)	SD(AN)	t value	p<
SDRR	59.68	35.2516	60.84	37.2861	0.0984	n.s.
MDRR	46.47	28.0106	48.00	29.3428	0.1640	n.s.
dSDRR	21.79	19.8846	21.42	23.2816	0.0525	n.s.
dMDRR	14.63	12.8072	14.58	15.3236	0.0115	n.s.
PNN50	4.11	4.5813	4.84	6.4314	0.4068	n.s.
RMSSD	27.89	22.3356	28.74	26.4067	0.1061	n.s.
LF	150.32	265.5330	109.16	121.4113	0.6145	n.s.
HF	26.74	45.2141	25.95	42.7271	0.0553	n.s.
LF/HF	18.39	43.5812	20.55	30.3078	0.1773	n.s.

The values of TRABI for the HRV indices from the comparative study in the morning and in the afternoon of the response to Valsalva manoeuvre and to rest in the patients with NMMI suggest that the parasympathetic sample leads to identical response in both time intervals (Figure 7.6). The mean value of the index is significantly lower than that in healthy individuals. *This result shows almost totally lost circadian nature of the vagal tone.* Only for the indice LF the value for TRABI is higher in the patients with NMMI compared to healthy individuals. The result supports the hypothesis presented above that the strongly decreased circadian nature of the vagus allows the manifestation of the circadian nature of the sympathicus.

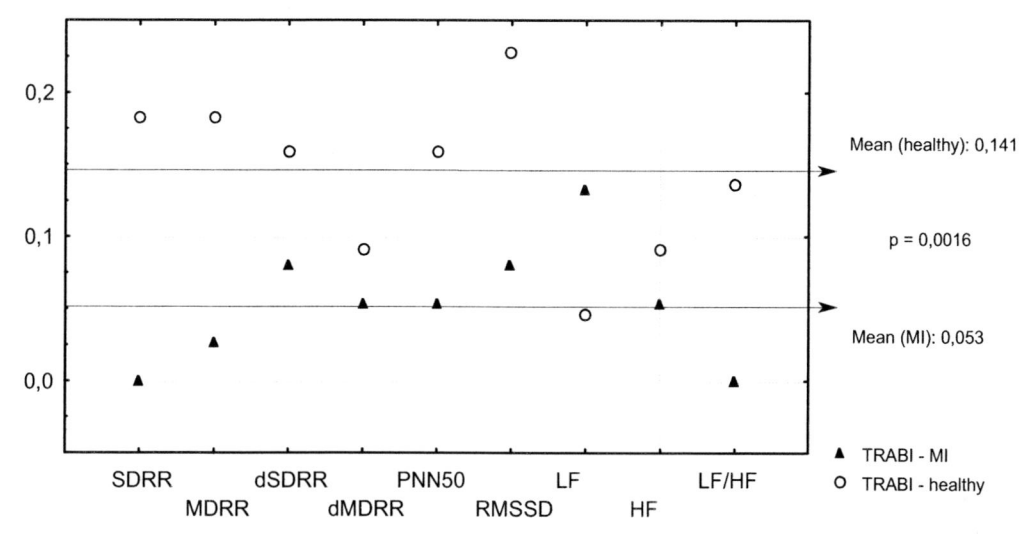

Figure 7.6. Values of TRABI for the HRV indices in healthy individuals and in patients with NMMI with comparison between morning and afternoon measurements during rest and with the Valsalva manouevre

Summarizing the results in patients with NMMI, we can assume that the circadian nature of the sympathetic part of the autonomic balance is preserved, with a ***strongly decreased circadian nature of the parasympathetic part of the balance.***

This result correlated with the results in patients with prolonged diabetes. These patients also lack a morning peak in the frequency of the MI incidents. The autonomic neuropathy has affected predominantly the vagal innervations. In long-term diabetics with infarction there is an abnormal change in the circadian nature of the autonomic balance.

It was already stressed that in healthy individuals the circadian nature of the parasympathicus prevails over that of the sympathicus. For the morning infarctions the mean value of TRABI is much higher in the comparative study between rest and with the handgrip test than the mean value in the comparative study during rest and with the Valsalva manoeuvre. It would be justifiable to conclude that the circadian nature of the sympathicus is more strongly pronounced. With the non-morning infarctions it is interesting to note that TRABI has very low mean values both in the first (0.082) and in the second study (0.053). This result shows strongly decreased circadian nature of the autonomic balance. The mean value of TRABI in the comparative study between rest and with the handgrip test remains higher than the mean value in the comparative study between rest and with the Valsalva manoeuvre, which shows that in the case of NMMI, too, there is abnormal circadian nature of the autonomic balance. For the sake of comparison, we shall recall that in the patients with unstable angina pectoris the circadian characteristic of the autonomic balance remains normal. The mean value of TRABI in the comparative study between rest and with the Valsalva manoeuvre in the patients with UA remains higher than the mean value of TRABI in the comparative study between rest and with the handgrip test. ***From these results we can*** *conclude that the sympathetic dysfunction is the trigger for the appearance and poor prognosis of cardiovascular incidents, whereas the vagal activity has a protective role.* The preserved circadian characteristic of the autonomic balance probably also has a protective function. As was indicated, both in the morning and in the non-morning MI, the circadian characteristic of the autonomic balance is abnormal, whereas the circadian nature of the balance is normal in the group of patients with unstable angina pectoris with good long-term prognosis.

REFERENCES

[1] Farrel, TG; Bashir, Y; Cripps, T; Malik, M; Poloniecki, J; Bennett, ED; Ward, DE; Camm, AJ. Riskc stratification for arrhythmic events in postinfarction patients based on heart rate variability, ambulatory electrocardiographic variables and signal-averaged electrocardiogram. *JAm Coll Cardiol.* 1991, 18, 687-607.

[2] Kleiger, RE; Miler, GP; Bigger, JT; Moss, AJ. Decreased heart rate variability and its association with increased mortality after acute myocardial infarction. *Am J Cardiol*, 1987, 59, 256-262.

[3] Kleiger, RE; Miler, GP; Krone, RJ; Bigger, JT. The independence of cycle length variability and exercise testing on predicting mortality of patients surviving myocardial infarction. *Am J Cardiol*,1990; 65, 408-411.

[4] Bigger, JT; Fleiss, JL; Steiman, RC; Rolnitzki, LM; Kleiger, RE; Rottman, JN. Correlation among time and frequency-domain measures of heart period variability two weeks after acute myocardial infarction. *Am J Cardiol*, 1992, 69, 891-898.

[5] Bigger, JT; Fleiss, JL; Steiman, RC; Rolnitzki, LM; Kleiger, RE; Rottman, JN. Frequency-domain measures of heart period variability and mortality after myocardial infarction. *Circulation*, 1992, 85,164-171.

[6] Bigger, JT; Fleiss, JL; Steiman, RC; Rolnitzki, LM; Shneider, WJ; Stein, PK. RR Variability in Healthy, Middle- Aged persons Compared With Patients With Chronic Coronary Heart Disease or Recent Acute Myocardial Infarction. *Circulation*, 1995, 91, 1936-1943

[7] Rich, MW; Saini, JS; Kleiger, RE; Carney, RM; TeVelde, A; Freedland, KE. Correlation of heart rate variability with clinical and angiographic variables and late mortality after coronary angiography. *Am J Cardiol*, 1988, 62, 714-717.

[8] Wennerblom, B; Lurje, L; Tygesen, H; Vahisalo, R; Hjalmarson, A. Patients with uncomplicated coronary artery disease have reduced heart rate variability mainly affecting vagal tone. *Heart*, 2000, 83, 290-294.

[9] Tsuji, H; Larson, M; Venditti, F; Manders, ES; Evans, JC; Feldman, CL; Levy, D. Impact of Reduced Heart Rate variability on Risk for Cardiac Events. *Circulation*, 1996, 94, 2850-2855.

[10] Singh, N; Mironov, D; Armstrong, P; Ross, AM; Langer, A. Heart Rate Variability Assessment Early After Acute Myocardial Infarction. *Circulation*, 1996, 93, 1388-1395.

[11] Wennerblom, B; Lurje, L; Solem, J; Tygesen, H; Uden, M; Vahisalo, R; Hjalmarson, Reduced heart rate variability in ischemic heart disease is only partially caused by ischemia. An HRV study before and after PTCA. *Cardiology*, 2000, 94, 146-151.

[12] Lanza, GA; Pedrotti, P; Pasceri, V; Lucente, M; Crea, F; Maseri, A. Autonomic changes association with spontaneous coronary spasm in patients with variant angina. *J Am Coll, Cardiology*, 1996, 5, 1249-1256.

[13] Lanza, GA; Pedrotti, P; Rebuzzi, AG; Pasceri, V; Qaranta, G; Maseri, A. Usefulness of the addition of heart rate variability to Holter monitoring in predicting in-hospital cardiac events in patients with unstable angina pectoris. *Am J Cardiol*, 1997, 80, 263-267.

[14] Singh, RB; Katrik, C; Otsuka, K; Pella, D; Pella, J. Coezyme Q (CoQ10). *Reasearch Biomed Pharmacother*, 2002, 56, Suppl. 2, 257-265.

[15] Huikuri, HV; Jokinen, V; Syvanne, M; Nieminen, MS; Airaksinen, J; Icaheimo, MJ; Koisyinen, JM; Kauma, H; Kesaniemi, AY; Majahalme, S; Niemela, KO; Frick, H. Heart Rate Variability and Progression of Coronary Atherosclerosis. *Arteriosclerosis, Thrombosis and Vascular Biology*, 1999, 19, 1979-1985.

[16] Krantz, D; Kop, W; Gabbay, F; Rozanski, A; barnard, M;Klein, J; Pardo, Y; Gottdiener, J. Circadian Variation of Ambulatory Myocardial Ischemia, *Circulation*, 1996, 93, 1364-1371.

[17] Deedwania, PC; Nelson, JR. Pathophysiology of silent myocardial ischemia during daily life: hemodynamic evaluation by simultaneous electrocardiographic and blood pressure monitiring. *Circulation*, 1990, 82, 1296-1304.

[18] Parker, JD; Testa, MA; Jimenez, AH; Tofler, GH; Muller, JE; Parker, JO; Stone, PH. Morning increase in ambulatory ischemia in patients with stable coronary artery disease. Importance of physical activity and increased cardiac demand. *Circulation*, 1994, 89, 604-614.

[19] Marchant, B; Stevenson, R; Vaishnav, S; Wilkinson, P; Ranjadayalan, K; Timmis, AD. Influence of the autonomic nervous system on circadian patterns of myocardial ischemia: comparison of stable angina with early postinfarction period. *Br Heart J*, 1994, 71, 329-333.

[20] Wennerblom, B; Lurje, L; Karlsson, T; Tygesen, H; Vahisalo, R; Hjalmarson, A. Circadian variation of heart rate variability and rate of autonomic change in the morning hours in healthy subjects and angina patients. *Int J Cardiol*, 2001, 79, 61-69.

[21] Huikuri,HV; Niemela, MJ; Ojala, S; Rantala, A; Ikaheimo, MJ; Airaksinen, KE. Circadian rhythms of frequency-domain measures of heart rate variability in healthy subjects and patients with coronary artery disease. Effects of arousal and upright posture. *Circulation*, 1994,90,121-126.

[22] Figueras, J;Lidon, RM. Early Morning Reduction in Ischemic Threshold in Patients With Unstable Angina and Significant Coronary Disease. *Circulation*, 1995, 92, 1737-1742.

[23] Cannon, CP; McCabe, CH; Stone, PH; Schactman, M; Thompson, B; Theroux, P; Gibson, T; Feldman, T; Kleiman, NS; Tofler, GH; Muller, JE; Chaitman, BR; Braunwold, E. Circadian variation in the onset of unstable angina and non-Q wave acute myocardial infarction (the TIMI III Registry and TIMI IIIB). *Am J Cardiol*, 1997, 79, 253-258.

[24] Lanza, GA; Patti, G; Pasceri, V; Manolfi, M; Sestito, A; Lucente, M; Crea, F; Maseri, Al. Circadian distribution of ischemic attacks and ischemia-related ventricular arrhythmias in patients with variant angina. *Cardiologia*, 1999, 44, 913-919.

[25] Hassan El-Tamimi; Mansour, M; Pepin, CJ; Wargovich, TJ; Chen, H. Circadian variation in Coronary Tone in Patients With Stable Angina. *Circulacion*, 1995; 92: 3201-3205.

[26] Uren,NG; Crake, T; Tousoulis, D; Seydoux, C; Davies, G; Maseri,A. Impairment of the myocardial vasomotor response to cold pressor stress in collateral dependent myocardium. *Heart*, 78, 61-67.

[27] Shaw, JA; Chin-Dusting, JPE; Kingwell, B; Dart AM. Diurnal Variation in Endothelium –Dependent Vasodilatation Is Not Apparent in Coronary Disease. *Circulation*, 2001, 103, 806-812.

[28] Tofler, GH; Brezinski, D; Shafer, AI; Czeisler, CA; Rutherford, JD; Willich, SN; Gleason, RE; Williams, GH; Muller, JE. Concurrent morning increase in platelet aggregability and the risk of myocardial infarction and sudden cardiac death. *New Engl J Med*, 1987, 316, 1514-1518.

[29] Brezinski, D; Tofler, GH; Muller, JE; Pohjola-Sintonen, S; Willich, SN; Schafer, AI; Czeisler, CA; Williams, GH. Morning Increase in Platelet Aggregability. Association With Assumption of the Upright Posture. *Circulation*, 1988, 78, 35-40.

[30] Ehrly, AM; Jung, G. Circadian rhythmof human blood viscosity. *Biorheology*, 1973, 10, 577-583.

[31] Bridges, AB; Scott, NA; McNeill, GP; Pringle, HT; Belch, JJF. Circadian variation of white blood cell aggregation and free radicals indices in men with ischemic heart disease. *Eur Heart J*, 1992, 13, 1632-1636.

[32] Masuda, T; Ogava, H; Miyao, Y; Yu, Q; Misumi, I; Sakamoto, T; Okubo, H; Okumura, K; Yasue, H. Circadian variation in fibrinolytic activity in patients with variant angina. *Br Heart J*, 1994, 71, 156-161.

[33] Manfredini, R; Galerani, M; Portalupi, F; Salmi, R; Chierci, F; Tassi, A; Rizzioli, E; Notarstefano, P; Risichella IS; Mirizio, AM; Fersini, C. Chronobiologic Aspects of Ischemic Coronary Artery Disease. *Jpn Heart J*, 1996, 37, 829-836.

[34] Ridker, PM; Manson, JA; Buring, JE; Muller, J; Hennenkens, C. Circadian Variation of Acute Myocardial Infarction and Effect of Low-Dose Aspirin in a Randomized Trial of Physicians. *Circulation*, 1990, 82, 897-901.

[35] Muller, JE; Stone, PH; Zoltan, MD; Rutherford, JD; Czeisler, CA; Parker, C; Poole, WK; Passamani, Eugene; Roberts, R; Robertson, T; Sobel, BE; Willerson, JT; Braunwold, E; and the MILIS Study Group. Circadian Variation in the Frequency of Onset of Acute Myocardial Infarction. *New Engl J Med*, 1985, 313, 1315-1321.

[36] Willich, SN; Linderer, T; Wegscheider, K; Leizorovicz, A; Alamercery, I; Schroder, R; and the ISAM Study Group. Increased Morning Incidence of Myocardial Infarction in the ISAM Study: Absence With Prior Beta-Adrenergic Blockade. *Circulation*, 1989, 80, 853-858.

[37] Goldberg, R; Bradi, P; Muller, J; Chen, Z; Groot, M; Zonnenveld, P; Dalen, JE. Time of Onset of Symptoms of Acute Myocardial Infarction. *Am J Cardiol*, 1990, 66, 140-144.

[38] Rana, JS; Mukamal, KJ; Morgan, JP; Muller, JE; Mittleman, MA. Circadian Variation in the Onset of Myocardial Infarction. *Diabetes*, 2003, 52, 1464-1468.

[39] Malik, M; Farrell, T; Camm, J. Circadian rhythm of heart rate variability after acute myocardial infarction and its influence on the prognostic value of heart rate variability. *Am J Cardiol*, 1990, 66,1049-1054.

Chapter 8

HEART FAILURE AND HEART AUTONOMIC BALANCE: CORRELATING CHANGES IN AUTONOMIC BALANCE CIRCADIAN CHARACTERISTICS AND VENTRICULAR ARRHYTHMIAS

Rada Prokopova
St. Anna University Hospital, Sofia, Bulgaria
Mikhail Matveev
Centre of Biomedical Engineering, Bulgarian Academy of Sciences, Bulgaria

HEART FAILURE AND HEART AUTONOMIC BALANCE

The activation of the sympathetic nervous system is one of the most serious pathophysiological dysfunctions in patients with chronic heart failure (CHF).

The level of the circulating catecholamines rises in such patients proportionally to the severity of the disease [1]. With the rise in the adrenalin level from 600 pg/ml to 900 pg/ml, the all-cause mortality increases from 48% to 80% [2].

Catecholamines are a major factor for the adaptation of the organism against the reduced stroke volume and hypotension, which cause hypoperfusion of vitally important organs. In the case of HF, catecholamines are also nonphysiologically distributed. They are more in the peripheral circulation than in the heart. Catecholamines are the cause for low stroke volume as a result of the compensatory tachycardia provoked by them. They also induce peripheral vasospasm, a rise in the peripheral vascular resistance with increased cardiac afterloading, poor tissue respiration and numerous metabolic disorders. Other pathophysiological disorders in CHF comprise the defect in the synthesis of catecholamines, the low degree of regulation of the beta-receptors (down regulation) and the change in the ratio between the beta-1- and beta-2-receptors. In the case of down regulation, the beta-receptors respond with difficulty to the catecholamine stimuli: they are "dazed" by the hypercatecholaminaemia. HF is also accompanied by a rise in the number of the alpha-receptors in the heart. Normally they prevail in the peripheral vessels. In CHF, the catecholamine stimulation of the alpha-receptors practically does not affect myocardial contractility, but it causes a strong constriction of the arterioles. At renal level this results in stimulation of the renin excretion and activation of RAAS.

The least favorable effects of the elevated level of catecholamines in the case of HF are: (i) intensification of the hypertrophy, (ii) aggravation of the ischaemia and activation of the necrosis of the myocytes, (iii) appearance of arrhythmias, (iv) increase of the heart rate.

The autonomic dysfunction is associated with the high mortality in patients with CHF. Therefore, large prospective studies assess HRV as a possible independent predictor for the onset of SCD in the case of HF. It has been found [3, 4] that HRV in HF is reduced. The indices SDRR, RMSSD and PNN50 have lower values, but the mechanism of the reduction is different. The indices PNN50 and RMSSD are influenced predominantly by the changes in the vagal activity, and they reflect the lowering of the parasympathetic tone. The decrease in SDRR is due to a complex mechanism: dysfunction (abnormalities) of the sympathicus, parasympathicus, the renin-angiotensin system, the haemoreceptor dysfunction, changes in the respiration and physical inactivity. The indice SDRR is a significant predictor of all-cause mortality. However, its highest predictive value is for death caused by HF progression. The low SDRR values reflect persisting neuroendocrine dysfunction, i.e., reflex activation of the renin-angiotensin-aldosterone and of the sympathetic systems. The abnormal activity of these systems leads to remodeling of the ventricles, and hence also to HF progression.

In [5], the authors have found that time- and frequency-domain HRV indices have a prognostic significance in patients with CHF both for identification of the patients with increased risk of all-cause mortality and for sudden cardiac death. And they have identified among the time-damain indices the significant predictive importance of the changes in the SDRR values both for the all-cause mortality and for sudden death. Among the spectral components, the reduction of LF during the day is a significant and independent predictor of SCD. In another study [6], HRV is investigated in patients with CHF only with frequency indices. The plasma noradrenalin is measured and the severity of the ventricular arrhythmias is assessed. The role of LF as independent predictor of sudden, probably rhythm death, is confirmed. The level of the plasma noradrenalin is higher in patients with significant LF reduction. This ostensibly paradoxical result (hypersumpatheticotony is usually associated in the literature with increase of LF) is explained with the reduced ability of the sinus node to respond to the extremely elevated sympathetic levels in the case of heart failure. We shall point out that we gave such an explanation of the reduced values of LF in 2002 in a study on hypertensive patients [7].

It was noted in Chapter 3 that VLF is influenced by thermoregulation, RAAS, parasympathetic activation and physical inactivity. The predictive value of the frequency-domain indices and their connection with the different causes of death in the case of CHF is studied using spectral analysis of HRV in [8]. The patients who died of progressive pump failure had reduced VLF values during the night. The authors argue that if in the case of CHF the low physical activity influences substantially the VLF values, it is to be expected that in gravely ill patients (due to their total immobilization) the daytime values would be most informative, which is not confirmed by the results. It seems more probable the changes in VLF to result from the activation of RAAS (see Chapter 3). It is confirmed that the reduced LF values reflecting the hypersumpatheticotony have a well manifested link with SCD.

The link between the hypersumpatheticotony and the appearance of ventricular arrhythmias (arrhytmogenesis) is well studied, although there also exist studies in which the sympathicoadrenergic mechanism of the arrhythmias is not unambiguously accepted (see, e.g., [9]). However, most authors confirm the leading role of the sympathetic activation as a triggering factor of the arrhythmias in pathological states: ischaemia, myocardial infarction, reperfusion and heart failure. In the case of CHF, the high level of the circulating catecholamines, as well as the elevated internal cardiac release of noradrenalin, stimulate the sympathetic nerve traffic and lead to higher risk of fatal arrhythmias and SCD. There are other arguments as well in support of the arrhythmogenicity of hypersumpatheticotony (see Chapter 2): (i) the possibility psychological stress to result in the appearance of ventricular tachycardias; (ii) stimulation of the sympathetic nerves (especially of the left cervicothoracic stellate ganglion) induces ventricular arrhythmias under suitable conditions: ischaemia, infarction and heart failure. There is evidence that in the case of infarction and heart failure, sympathetic stimulation induces a heterogeneous and not a homogeneous adrenergic stimulation, which is more arrhythmogenic [10]. Other authors [11] have found that complex ventricular extrasystoles appear in patients subjected to a two-minute balloon occlusion of the stenotic coronary artery, who have a significant decrease of HRV. In the patients with increase in the HRV, complex ventricular rhythmic disorders are lacking. The link between the absence of complex ventricular arrhythmias and the higher HRV is probably determined by the parasympathetic activation, which has a protective role and reduces the vulnerability of the myocardium.

Other researchers also confirm the link between the LF reduction, used as marker of sympathetic activity and the start of ventricular tachycardias. It has been proven [12] that the prevalent number of patients with CHF demonstrate reduced LF before the onset of ventricular tachycardias; fewer patients demonstrate an increase of LF. The cited authors associate that opposite reaction with the different basal sympathetic level in both groups of patients.

It should be accepted that the link between hypersumpatheticotony and/or lowered parasympathetic tone and elevated risk of ventricular rhythmic disorders has been proven. In Chapter Four, when we discussed the circadian nature of the autonomic balance, we pointed out that in the morning hours the balance is with the highest level of the sympathicus and with the lowest level of the parasympathicus. The circadian characteristic of the autonomic balance correlates with the circadian characteristic of catecholamines. Healthy subjects maintain a high parasympathetic and a low sympathetic level during the night. The circadian characteristic usually changes in the different diseases of the cardiovascular system. The ventricular rhythmic disorders also have a circadian characteristic with a peak in the morning and a second smaller peak in the afternoon [13]. This circadian nature corresponds to the higher incidence of the cardiovascular incidents and SCD in the morning.

The autonomic dysfunction plays an important role in the pathophysiology of SCD and the progression of CHF. Due to the circadian peak at the onset of SCD and the ventricular rhythmic disorders in the morning, it is logical to study the circadian characteristics of the autonomic balance in the case of CHF. We shall note that such studies are almost lacking. In [14] the authors report that the circadian nature of HRV is preserved in the case of CHF. In another study [15], the patients with heart failure are divided into two groups: one group comprises the persons with preserved physiological circadian nature of the HRV indices reflecting predominantly the vagal activity: RMSSD and HF. Their values are the highest during the night, and diminish in the morning hours. The patients in this group lack a peak of the ventricular tachyarrhythmias in the morning. The other group includes patients with inversion of the circadian nature for the indices RMSSD and HF. These patients have a morning peak of the ventricular tachyarrhythmias.

Restoration of the autonomic balance in patients with CHF could be expected by impacting the hypersumpatheticotony or the reduced parasympathetic tone with suitable drugs. Both vagomimetics and adrenergic blockers improve the autonomic balance in the patients with CHF [16, 17]. It would be justifiable to expect that medical intervention correcting the main pathophysiological disorder in the case of CHF would limit the progression of heart failure and the risk of sudden cardiac death.

Tjeerdsma, Szabo and van Wijk [18] have studied the autonomic balance through HRV and functional tests influencing the nervous system, as well as the plasma catecholamine level in patients with CHF before and after treatment with Methoprolol. There are no significant differences in the catecholamine levels and in the results of the functional tests. However, after treatment with Methoprolol, although the change in the values of SDRR and the total spectral power is not significant, there is a tendency towards restoration of the autonomic balance: the parasympathetic effects increase and the sympathetic ones decrease. In a study comprising 1,094 patients with CHF [19] it is reported that the patients treated with the alpha-1- and beta-receptor blocker Carvedilol have a lower mortality from aggravation of the heart failure and SCD. Boudonas, Psirropoulos, Ginopoulos, et al. [20] have studied the effect of Carvedilol on HRV, the dispersion of the Q-T interval (QTd) (indice for homogeneity of the ventricular repolarization) and the ventricular tachycardias in patients with CHF. The results suggest that Carvedilol increases the values of HF, reduces the LF/HF ratio and QTd, and lowers the class of the ventricular tachyarrhythmias. It may be assumed that the normalization of the autonomic balance and homogenization of the ventricular repolarization correlate with the lowered class of the ventricular tachyarrhythmias. Suitable treatment reduces the vulnerability of the myocardium; beta-blocker therapy protects against SCD. In [21] the authors have studied the effect of Carvedilol on the improvement of the autonomic balance in patients with heart failure. They have found increase of the values of the indices for HRV in the time and frequency area as a reflection of the improved autonomic regulation of the heart.

The cited survey leads to the conclusion that the restoration of the autonomic balance is a major factor for prevention of the progression of CHF and SCD. The drug therapy normalizing the autonomic balance ought to be linked with the specific circadian characteristic of the cardiovascular incidents on account of their high risk in patients with heart failure. We carried out a special study on the circadian nature of the autonomic balance of the heart in patients with CHF and on the influence of the circadian nature on the cardiac risk before and after treatment [21].

CORRELATING CHANGES IN HEART AUTONOMIC BALANCE AND VENTRICULAR ARRHYTHMIAS REFLECTING THE POSITIVE EFFECT OF HEART FAILURE TREATMENT

We studied 40 patients with CHF with ischaemic cardiomyopathy in sinus rhythm and ejection fraction (EF) < 40%. The alpha-1- and beta-2-receptor blocker Carvedilol (Dilatrend ®) was administered for three months in gradually increasing doses from 3.125 to 50 mg. Before and during the administration of Carvedilol, the patients were subjected to standard therapy with diuretic and ACE inhibitor without additional antiarrhythmic therapy. The patients in the group are without diabetes, renal failure, neurological complaints and other severe diseases. The changes in the autonomic balance were traced with the indices for HRV and the TRABI in 20 of the treated patients. These patients lacked extrasystolic arrhythmia in the morning and in the afternoon ECG records intended for analysis of HRV. The values of TRABI for each HRV indice before and after treatment in patients with CHF and in healthy subjects are given in Table 8.1.

Table 8. 1. Values of TRABI for HRV indices in patients with CHF before and after treatment and in healthy persons.

HRV indices	Before treatment RS vs. HG	Before treatment RS vs. VM	After treatment RS vs. HG	After treatment RS vs. VM	Healthy subjects RS vs. HG	Healthy subjects RS vs. VM
SDRR	0.600	0.099	0.000	0.000	0.045	0.182
MDRR	0.391	0.299	0.000	0.099	0.068	0.182
dSDRR	0.391	0.099	0.099	0.099	0.091	0.159
dMDRR	0.099	0.201	0.299	0.099	0.114	0.091
PNN50	0.099	0.099	0.299	0.299	0.159	0.159
RMSSD	0.099	0.201	0.099	0.201	0.159	0.227
VLF	0.201		0.000		0.091	
LF	0.391	0.201	0.000	0.391	0.045	0.045
HF	0.099	0.386	0.201	0.201	0.000	0.091
LF/HF	0.391	0.386	0.000	0.201	0.227	0.136

The severity of the ventricular arrhythmia before and after the treatment is assessed using 24-hour ECG-recordings.

Characteristics of the Autonomic Balance in Patients before Treatment with Carvedilol

In the patients with CHF before treatment, the values of TRABI for the indices for HRV in a comparison between the response to stimulation with the handgrip test and rest in the morning and in the afternoon (Table 8.1 and Figure 8.1) manifest a marked circadian nature in the autonomic balance – to a degree that is not registered in healthy subjects. The values of the indicator for the indices SDRR, MDRR, dSDRR, LF, LF/HF are close to 0.4, which indicates a substantial change in the afternoon study, compared to the results of the morning study. TRABI has the highest value for the indice SDRR: 0.6. For the sake of comparison, in healthy subjects the highest value of the index is 0.227 for LF/HF.

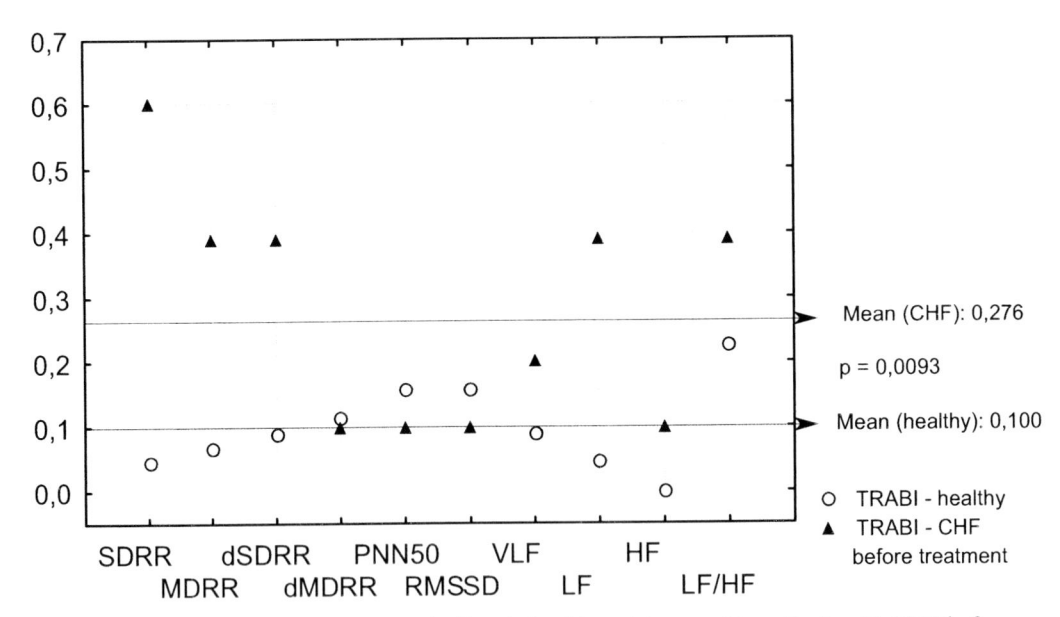

Figure 8.1. Values of TRABI for the HRV indices in healthy subjects and in patients with CHF before treatment in a comparison between morning and afternoon measurements during rest and with the handgrip test.

Consequently, in the case of CHF there is a very pronounced change in the autonomic balance in the different time intervals. In view of the character of the handgrip test – predominantly sympathetic stimulation – it may be claimed that stimulation of the sympathetic part induces a different response in intensity during the different time intervals in the case of CHF. The indices PNN50 and RMSSD, which reflect the state of the parasympathetic part, have lower values of the index compared to the values in healthy individuals. We can consider that the parasympathetic tone does not change substantially during the different hours of the day and this change is only relative with respect to the change in the sympathetic tone.

The cited specificities find support in the analysis of the statistical evaluations of the actual indices for HRV in the group of patients with CHF. The mean values of the indices during rest in the morning and in the afternoon (Table 8.2) do not differ significantly for most of the indices. Their values are lower than the values in healthy subjects – see Table 6.2.

As we pointed out already, in healthy individuals there is no significant difference between the values of the indices with the handgrip test in the morning and in the afternoon (Table 6.3). In patients with HF, the comparison between the values in the morning and in the afternoon with the handgrip test revealed a significant difference for all indices except VLF (Table 8.3). The values are lower in the afternoon hours. A possible explanation of this fact is that the sympathetic hyperactivation, which reduces the capacity for response to stress in the morning hours, allows response by the vagus. Only the response of the lower parasympathetic tone appears clearly in the afternoon hours.

Comparison between the values of the indices SDRR, MDRR, dSDRR and LF during rest and during the handgrip test in the morning (Tables 8.2 and 8.3) reveals that the sympathetic stimulation does not cause any reaction in these hours. This is explained with the impossibility to stimulate the sympathicus additionally due to the high hypersumpatheticotony level. Comparison between the values of the same indices during rest and during the handgrip test in the afternoon reveals a considerable difference. In this case there is already a possibility for additional stimulation, due to the lower tone of the sympathicus in the afternoon.

Table 8.2. Values of the HRV indices in the morning (M) and afternoon (AN) during rest before the treatment.

	Mean(M)	SD(M)	Mean(AN)	SD(AN)	t value	$p <$
SDRR	26.80	17.3118	21.80	15.1888	1.3731	n.s.
MDRR	21.40	15.4855	17.40	13.1833	1.2439	n.s.
dSDRR	12.40	10.0399	9.20	8.2583	1.5568	n.s.
dMDRR	8.00	5.7009	6.00	5.3385	1.6196	n.s.
PNN50	1.40	2.1909	1.00	1.7321	0.9058	n.s.
RMSSD	16.80	11.7132	12.20	10.0846	1.8823	n.s.
VLF	142.60	159.8399	114.20	129.1654	0.8740	n.s.
LF	74.40	75.5401	34.80	25.6164	3.1399	0.05
HF	25.80	27.9589	14.60	18.9552	2.0970	0.05
LF/HF	3.26	1.6431	3.56	2.3581	0.6672	n.s.

The values of TRABI for the HRV indices, when the reaction to stimulation with the Valsalva manoeuvre is compared with rest in the morning and in the afternoon (Table 8.1 and Figure 8.2), do not differ from the values of the index in healthy subjects to such a degree as with the handgrip test. For SDRR, RMSSD and especially for PNN50, the values of the index are lower in the patients with CHF compared to the values in healthy individuals. There is a slight change in the results of the morning and afternoon study. This is evidence of a less manifested circadian nature in the parasympathetic part of the autonomic balance. As in this case the evaluations for TRABI are obtained with predominantly vagal stimulation, the low values for the parasympathetic indices indicate total absence of change in the parasympathetic tone in both time intervals. This conclusion is also supported by the above-mentioned only relative change compared to the sympathetic tone. For the frequency indices LF, HF, LF/HF, the values of the index are higher than the values in healthy individuals. We explain it with the mixed character of the Valsalva manoeuvre. Probably in the case of heart failure – due to the abnormally strong reaction of the sympathicus in the morning – the results of the test reveal more prominently the sympathetic part of the stimulation and the values of the index indicate a more marked time-related change.

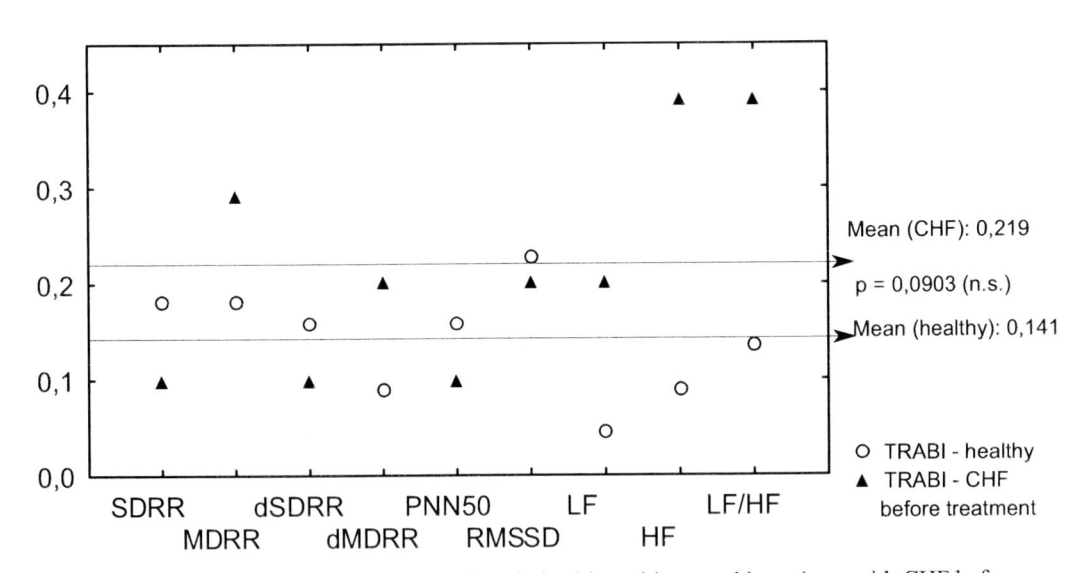

Figure 8.2. Values of TRABI for the HRV indices in healthy subjects and in patients with CHF before treatment, with comparison between morning and afternoon measurements during rest and with the Valsalva manoeuvre.

Table 8.3. Values of the HRV indices in the morning (M) and afternoon (AN) during the handgrip test before the treatment.

	Mean(M)	SD(M)	Mean(AN)	SD(AN)	t value	$p<$
SDRR	27.20	18.1301	16.80	8.5849	3.2789	0.05
MDRR	22.20	15.6109	13.40	6.5038	3.2910	0.05
dSDRR	13.80	14.2724	4.60	2.4083	4.0200	0.05
dMDRR	9.60	9.9900	3.40	1.6733	3.8712	0.05
PNN50	1.80	4.0249	0.00	0.0000	2.8284	0.05
RMSSD	18.40	17.1988	7.20	3.2711	4.0461	0.05
VLF	31.20	32.9803	21.80	23.9625	1.4583	n.s.
LF	54.60	54.2614	7.00	5.7879	5.5168	0.05
HF	10.40	12.5817	1.40	2.6077	4.4300	0.05
LF/HF	5.65	2.7224	14.37	17.1609	3.1726	0.05

As we pointed out, the comparison of the statistical evaluations of the indices shows almost no significant difference between their mean values during rest in the morning and in the afternoon (with the exception of the two frequency indices LF and HF). Conversely, during the Valsalva manoeuvre (Table 8.4), all indices with the exception of these two frequency indices have significantly differing values in the two studies. In the afternoon hours the values are significantly lower with the Valsalva manoeuvre. It is possible that in the morning hours the high sympathetic tone, but with blocked capacity for response to stimulation, allows response of the vagus to the parasympathetic test, whereas in the afternoon hours only the reduced parasympathetic activity is manifested. The mean values of the frequency indices follow the same tendency of change, albeit insignificantly.

This result again indicates that upon stimulation the frequency-domain indices are extremely sensitive and do not reflect well the momentary state of the autonomic balance. Nevertheless, the comparison between the values of the indices during rest and during the

Valsalva manoeuvre in the morning shows that there exists a reaction to the stimulation, i.e., in the morning the level of the parasympathicus tends to be lower and allows additional stimulation. The afternoon measurements reveal almost no difference between the values of the indices during rest and during the test. This result reflects the relative stabilization of the parasympathicus in the afternoon, but at a lower level than the level in healthy individuals.

Table 8. 4. Values of the HRV indices in the morning (M) and afternoon (AN) with the Valsalva manoeuvre before the treatment.

	Mean(M)	SD(M)	Mean(AN)	SD(AN)	*t* value	*p* <
RRSD	44.60	37.9776	28.20	29.6934	2.1516	0.05
RRMD	37.20	32.3914	21.60	23.8390	2.4532	0.05
dRRSD	19.60	19.9575	8.60	5.5946	3.3565	0.05
dRRMD	14.20	14.8223	5.40	3.2094	3.6698	0.05
PNN50	4.00	5.8737	1.00	1.7321	3.0984	0.05
RMSSD	25.80	23.6897	11.40	7.2664	3.6754	0.05
LF	89.60	133.3334	93.60	177.9685	0.1138	n.s.
HF	13.80	20.7894	6.40	11.5672	1.9672	n.s.
LF/HF	13.33	9.2855	19.77	30.1357	1.2916	n.s.

Characteristics of the Autonomic Balance after the Treatment

Comparison between the data during rest and during the handgrip test in the morning and in the afternoon after treatment with Carvedilol shows that the values of TRABI for the indices SDRR, MDRR, dSDRR, LF and LF/HF are diminished and are even lower than the values in healthy individuals (Table 8.1 and Figure 8.3). This result demonstrates that the therapy leads to a change in the circadian characteristic of the autonomic balance. In view of the constellation of indices in which the time-domain indices reflect predominantly the general changes in HRV, and LF – the sympathetic tone and its participation in the autonomic balance, the hypothesis on the decrease of the sympathetic part in the circadian characteristic of vegetative control is justified. At the same time, the values of the index for the indices PNN50 and HF are elevated both with respect to the values before therapy, and with respect to the values in healthy subjects. As the handgrip test is predominantly a sympathetic test, the elevated values of the indices that are sensitive to changes in the parasympathetic activity reflect the relative prevalence of the vagal tone over the suppressed sympathetic activity in the autonomic balance.

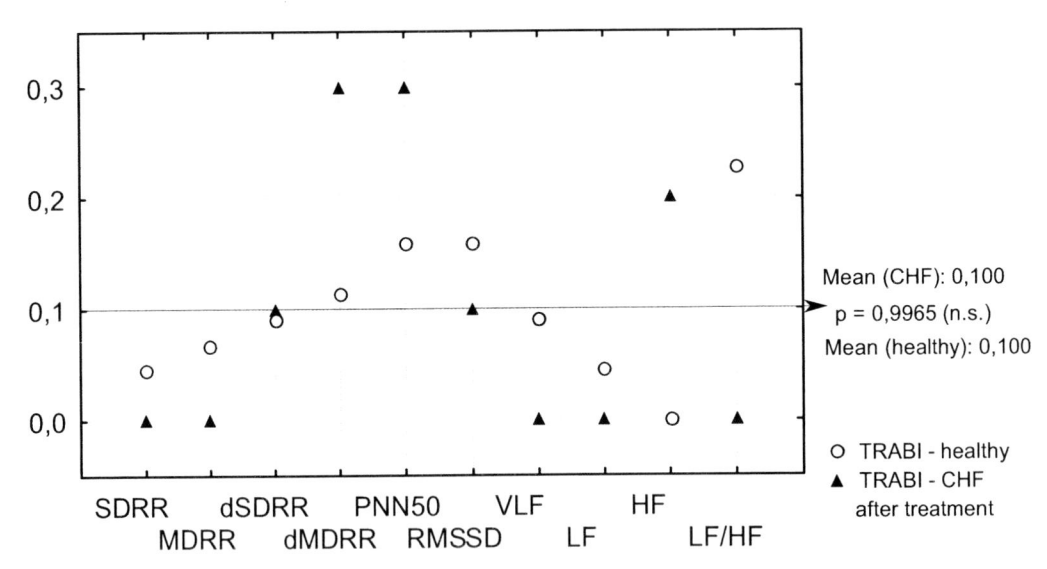

Figure 8.3. Values of TRABI for the HRV indices in healthy subjects and in patients with CHF after treatment, with comparison between morning and afternoon measurements during rest and during the handgrip test.

The comparison of the statistical evaluations of the indices for HRV before and after therapy during rest in the morning does not reveal differences, except in the case of HF whose value increases and VLF where it decreases (Table 8.2 and 8.5). During rest in the afternoon, the values of all indices rise after therapy, with the exception of VLF whose value diminishes.

Consequently, HRV in the case of CHF is not completely normalized after the therapy, but the rise in the values of the indices in the afternoon indicates partial restoration of the autonomic balance. The lowered VLF values both in the morning and in the afternoon after treatment is explained with the suppression of RAAS.

Table 8.5. Values of the HRV indices in the morning (M) and afternoon (AN) during rest after the treatment.

	Mean(M)	SD(M)	Mean(AN)	SD(AN)	t value	$p <$
SDRR	25.00	10.2713	38.80	25.2131	3.2058	0.05
MDRR	19.80	9.2304	32.40	24.2858	3.0672	0.05
dSDRR	12.20	11.2116	25.00	21.3073	3.3623	0.05
dMDRR	8.80	7.4632	18.60	15.9781	3.5146	0.05
PNN50	2.00	3.9370	16.80	29.2010	3.1767	0.05
RMSSD	18.60	16.1028	50.60	57.4134	3.3941	0.05
VLF	69.00	62.0081	43.40	24.3167	2.4309	0.05
LF	70.60	100.3559	71.40	79.7076	0.0395	n.s.
HF	54.20	86.8602	83.80	136.8565	1.1549	n.s.
LF/HF	3.12	4.8336	2.58	2.4256	0.6362	n.s.

After administration of Carvedilol, in the studies during rest in the morning and in the afternoon there are significant changes in the values of most indices, unlike the results in the analogous study before the treatment. In the afternoon studies, the values of the indices are significantly higher than those in the morning. However, since the parasympathetic influence prevails during rest, we can assume that the therapy with Carvedilol suppresses more strongly the vagal tone in the morning hours, which entails a more significant time change. This result also correlates with the higher mean value of TRABI with the Valsalva manoeuvre after treatment than the value for healthy individuals. There is an elevated circadian nature of the parasympathicus, which is probably due to the relatively lower vagal tone in the morning hours compared to that in healthy individuals.

The absence of significant differences for the frequency indices LF, HF and LF/HF probably stems from the suppression of the sympathicus after the therapy. The comparison between the values after treatment in the morning and in the afternoon with the handgrip test reveals that after the treatment the mean values of most of the indices do not differ significantly in the morning and in the afternoon. We shall recall that prior to treatment there is a significant difference between the mean values in the morning and in the afternoon for all indices, with the exception of VLF. In this way, the result after treatment is close to the result in healthy individuals. There is a significant difference between the mean values in the morning and in the afternoon with the handgrip test only for the indices PNN50 and HF (Table 8.6), their values being higher in the afternoon hours. This is additional confirmation for the possibility the parasympathetic activity to overcome the hypersumpatheticotony, at least in the afternoon hours, which was restored as a result of the therapy.

The comparison between the mean values of the indices for HRV, according to data from the measurements in the morning during rest and during the handgrip test after the therapy with Carvedilol, leads to a result that is contrary to the result of the analogous comparison before the treatment. After the therapy, there was a significant difference in all indices between the values during rest and during the handgrip test. Apparently, the sympathetic tone in the morning hours is reduced as a result of the treatment. This relatively lower level of the tone allows the sympathicus to respond to the stimulation.

Table 8.6. Values of the HRV indices in the morning (M) and afternoon (AN) during the handgrip test after the treatment.

	Mean(M)	SD(M)	Mean(AN)	SD(AN)	t value	$p<$
SDRR	31.60	19.9700	38.40	19.2951	1.5488	n.s.
MDRR	23.80	12.5579	32.20	18.6199	2.3655	0.05
dSDRR	25.40	37.4473	23.80	21.9249	0.2332	n.s.
dMDRR	21.00	33.1662	19.40	19.9324	0.2615	n.s.
PNN50	5.60	10.3102	14.40	25.0958	2.0514	0.05
RMSSD	34.40	45.1475	43.60	45.9870	0.9029	n.s.
VLF	43.00	40.8656	36.80	21.1234	0.8524	n.s.
LF	12.60	10.6442	26.00	14.4049	4.7317	0.05
HF	8.20	9.1488	43.00	60.5970	3.5914	0.05
LF/HF	5.74	7.6510	3.71	5.0352	1.3976	n.s.

In the afternoon there are less pronounced differences in the values of the indices for HRV. Apparently, stability of the sympathicus is attained after the therapy in the afternoon and its stimulation with the test is more difficult. Only with the frequency indices LF and HF there is a substantial decrease in the values during the handgrip test compared to the values during rest. Although in the afternoon the sympathicus responds less strongly to stimulation, nevertheless a higher than normal level of the sympathicotonia is to be expected in spite of the therapy. The frequency indices reflect this elevated tone during stimulation.

The value of TRABI according to data from the measurements during rest and during the Valsalva manoeuvre in the morning and in the afternoon after treatment for six of the indices decreases, and in one (RMSSD) does not change with respect to the evaluation prior to the treatment (Table 8.1 and Figure 8.4). This result shows that the therapy reduces the circadian nature of the parasympathetic part of the balance. The value of the index is higher only for the indices PNN50 and LF. In the case of PNN50, this reflects the relatively restored (with respect to the lower sympathetic level) circadian nature of the parasympathicus both in the morning and in the afternoon. The higher value of the index for LF reflects the result of the therapy with the alpha- beta-blocker Carvedilol. With blocked sympathicus, the changes in the LF values are reduced and this index can already reflect the time-related changes with greater sensitivity. The values of TRABI for HF and LF/HF are considerably reduced compared to the values before treatment as a reflection of the reduced circadian nature of the parasympathicus. The circadian nature of the parasympathicus, evaluated using TRABI, remains higher than the circadian nature in healthy individuals. The values of the "parasympathetic" indices PNN50 and HF are considerably higher than the values in healthy individuals. As will be pointed out below, this result confirms the stronger reduction of the vagal tone in the morning after the therapy, hence its more pronounced circadian characteristic.

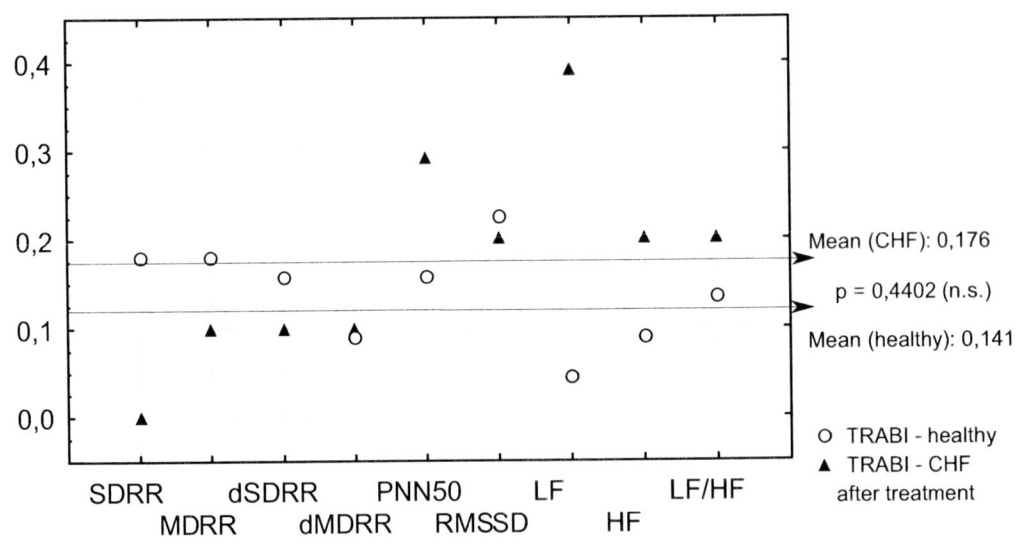

Figure 8.4. Values of TRABI for the HRV indices in healthy subjects and in patients with CHF after treatment, with comparison between morning and afternoon measurements during rest and during the Valsalva manoeuvre.

During the Valsalva manoeuvre, the mean values of all indices for HRV, according to data from the studies in the morning and in the afternoon after therapy, do not show significant differences (Table 8.7). This result is similar to the result for healthy individuals, who also lack significant differences in that study in the morning and in the afternoon. There is a tendency towards normalization of the parasympathetic part of the vegetative balance. The values of TRABI also point to such a tendency.

The mean values of all indices in the morning after the therapy, according to data from the study with the Valsalva manoeuvre, increase compared to their values according to the data during rest. For PNN50 and RMSSD the increase is considerable. The reason is in the elevated reactivity of the parasympathicus during stimulation in the morning hours as a result of the therapy. A part of the elevated reactivity is also due to the sympathetic tone that is suppressed after the therapy. The mean value of LF (as well as before treatment) decreases, and of LF/HF increases due to the mixed character of the test.

In the afternoon, the change in the mean values of the indices, according to data from the measurements during rest and during the Valsalva manoeuvre, is not considerable. Nevertheless, the increase in the mean values of the indices SDRR, MDRR, dSDRR, dMDRR during the Valsalva manoeuvre with respect to the measurement during rest indicates a tendency towards general restoration of HRV.

Table 8.7. Values of the HRV indices in the morning (M) and afternoon (AN) with the Valsalva manoeuvre after the treatment

	Mean(M)	SD(M)	Mean(AN)	SD(AN)	t-value	p<
SDRR	56.00	38.5227	53.20	27.0869	0.3760	n.s.
MDRR	47.60	36.2119	45.60	24.5418	0.2892	n.s.
dSDRR	29.80	34.0176	30.60	28.1745	0.1145	n.s.
dMDRR	23.60	30.1629	26.20	28.0571	0.3992	n.s.
PNN50	15.00	28.0268	11.60	16.9204	0.6568	n.s.
RMSSD	60.80	89.2340	47.60	45.2581	0.8344	n.s.
LF	40.00	30.6023	35.20	28.0393	0.7314	n.s.
HF	9.80	7.8230	9.20	8.4676	0.3292	n.s.
LF/HF	9.52	11.3220	6.98	6.7895	1.2168	n.s.

Considerably reduced mean values of HF during the Valsalva manoeuvre with respect to rest are observed both in the morning and in the afternoon. This result is difficult to comment, because it presupposes suppression of the parasympathetic tone during the stimulation. However, a similar effect is reported by other authors as well (see, e.g., [23]), explaining it with the alpha-blocking activity of Carvedilol.

Frequency of the Ventricular Rhythmic Disorders before and after Treatment of CHF

Table 8.8 shows the frequencies of the ventricular rhythmic disorders before and after the administration of Carvedilol. For every group of disorders the difference between the frequencies is highly significant ($p<0.001$).

Table 8.8. Frequency of the ventricular rhythmic disorders before and after treatment with Carvedilol.

	Before treatment	After treatment	$p <$
Polytopic ventricular extrasystoles	40 subjects (100.0%)	16 subjects (40.0%)	0.001
Grouped ventricular extrasystoles	36 subjects (90.0%)	6 subjects (15.0%)	0.001
Ventricular tachycardia	20 subjects (50.0%)	4 subjects (10.0%)	0.001

Before treatment, 63 % of the ventricular rhythmic disorders are concentrated between 6 and 11 a.m. During these hours, 78% of the dangerous ventricular arrhythmias (grouped ventricular extrasystoles and ventricular tachycardias) are registered. After Carvedilol treatment for three months, there is no significant difference between the frequency of the ventricular rhythmic disorders in these hours and in other intervals of the active period.

Conclusions

The discussions so far can be summarized with several conclusions concerning the state of the autonomic balance and its circadian nature before and after the therapy with Carvedilol:

1. Before the therapy, the variability of the cardiac rhythm is reduced considerably both in the morning and in the afternoon. The circadian characteristic of the autonomic balance, evaluated through TRABI, is strongly disturbed in persons with CHF. The values of the index during the handgrip test are high for some of the indices. The abnormality in the circadian characteristic concerns predominantly the sympathetic component in the balance. It is characteristic that with the sympathetic test the untreated persons respond in the morning with a strong rise of the sympathetic level. With vagal stimulation, TRABI has low values for most indices as a reflection of the lowered circadian nature of the parasympathetic component in the autonomic balance in patients with CHF. Parasympathetic stimulation shows suppressed reactivity of the parasympathicus irrespective of the hours of the day and night. The stimulation cannot overcome the existing hypersumpatheticotony.
2. After the therapy there is no tendency to increase the values of the indices for HRV in the morning to the values in healthy individuals. In the afternoon the values of the time-domain indices approach normal values. For the frequency indices this tendency is not marked due to the residual hypersumpatheticotony. After the therapy the evaluations of TRABI for the indices for HRV during the handgrip test in patients with CHF are lower than the evaluations in healthy subjects. The therapy overcomes the hypersumpatheticotony in the morning, but does not lead to restoration of the normal circadian characteristic of the autonomic balance. For the parasympathetic indices the evaluations of the index are higher. The parasympathicus becomes reactive due to the lower tone of the sympathicus. Upon stimulation with the Valsalva manoeuvre, the circadian nature of the parasympathetic part of the balance decreases, but remains higher than that in healthy individuals. The reason can be sought in the reduction of the parasympathetic tone in the morning hours with

Carvedilol therapy. Was we pointed out, this result is probably connected with its alpha-blocking activity.

Carvedilol also tends to have a negative influence on the restoration of the circadian nature of the parasympathetic tone, because it induces a more pronounced morning suppression of the parasympathicus.

On the whole, however, the therapy with Carvedilol influences the autonomic balance. It reduces the sympathetic tone (predominantly in the morning), but does not normalize the circadian characteristic of the balance. These changes are illustrated well with Figure 8.5, which gives the profiles of TRABI values from the comparative studies during rest and during the handgrip test before and after the treatment. There is a significant difference between the mean values of the index for the HRV indices. The parasympathetic part of the autonomic balance is slightly influenced, whereby the perceptible tendency towards higher parasympathetic tone with stimulation in the morning is rather relative against the background of the suppressed sympathetic tone. The changes in the autonomic balance after treatment compared to its state before the administration of Carvedilol can be traced on Figure 8.6. However, in this case the mean values of the index for the HRV indices from the comparative studies during rest and during the Valsalva manoeuvre do not differ significantly. As can be seen from the values of the index, the circadian nature of the autonomic balance in patients with HF is abnormal prior to the treatment. TRABI with the handgrip test has a higher value than the value of the index with the Valsalva manoeuvre, i.e., the circadian nature of the sympathicus is more pronounced. After the treatment, the circadian profile of the autonomic balance is normalized, whereby the value of TRABI for the Valsalva manoeuvre becomes higher than the value of the index with the handgrip test. However, the value of the index with the Valsalva manoeuvre after treatment remains higher than the value in healthy individuals. The reason for this is the stronger suppression of the parasympathetic tone in the morning hours with Carvedilol therapy, compared to the parasympathetic tone in healthy individuals.

3. The partial restoration of the autonomic balance after treatment with Carvedilol reduces the frequency of the ventricular arrhythmias throughout the entire 24-hour period. Even during the high risk morning hours under stress the frequency of the ventricular rhythmic disorders is lower.

The sympathetic tone, reduced as a result of the therapy, the suppressed stress response in the morning, as well as the slight rise in the parasympathetic tone, have a preventive effect on the risk of severe ventricular arrhythmias, and consequently also on SCD.

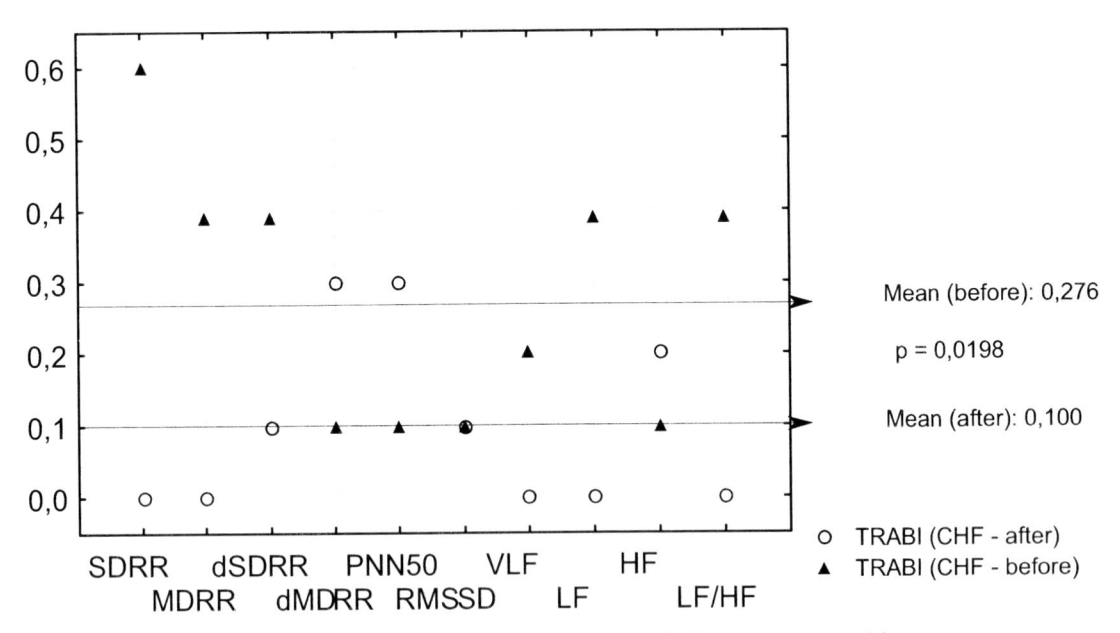

Figure 8.5. Values of TRABI for the HRV indices before and after treatment, with comparison between morning and afternoon measurements during rest and during the handgrip test.

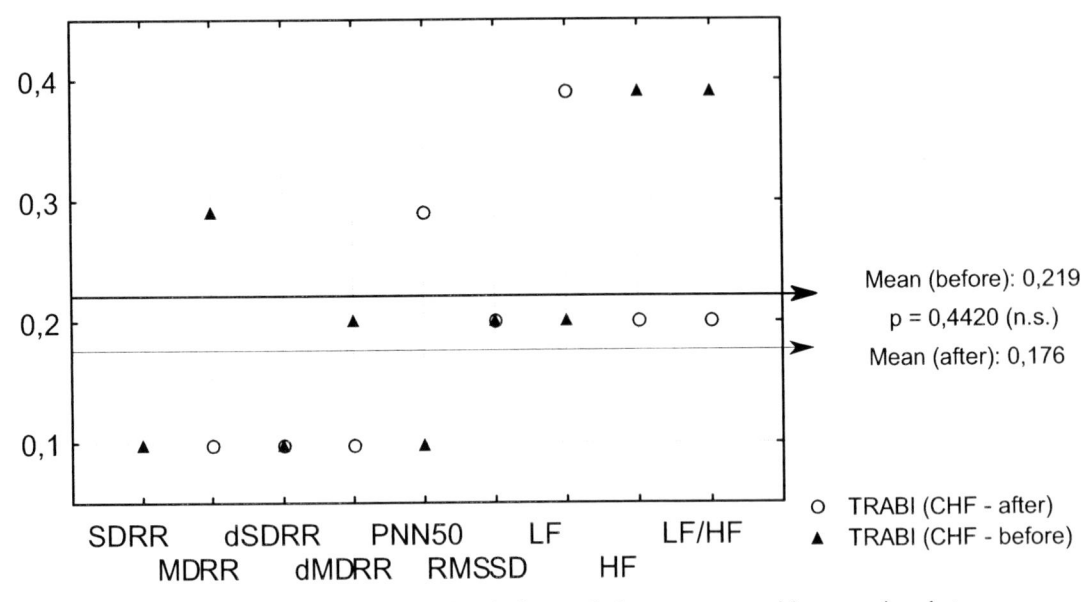

Figure 8.6. Values of TRABI for the HRV indices before and after treatment, with comparison between morning and afternoon measurements during rest and during the Valsalva manoeuvre.

REFERENCES

[1] Thomas, JA; Marks, BH. Plasma norepinephrine in congestive heart failure. *Am J Cardiol*, 1978, 41, 233-243.

[2] Bolte, H. *Katecholamine and Vasodilatantien bei Herzinsuffizienz*. Berlin – Heidelberg – New York: Springer Verlag; 1981.

[3] Nolan, J; Batin, PD; Andrews, R; Lindsay, SJ; Brooksby, P; Mullen, M; Baig, W; Flapan, AD; Cowley, A; Prescott, RJ; Neilson, JMM; Fox, KAA. Prospective Study of Heart Rate Variability and Mortality in Chronic Heart Failure. *Circulation*, 1998, 98,1510-1516.

[4] Bonaduce, D; Petretta, M; Marciano, F; Vicario, MLE; Apicella, C; Rao, MAE; Nicolai, E; Volve, M. Independent and incremental prognostic value of heart rate variability in patients with chronic heart failure. *Am Heart J*, 1999, 138, 273-284.

[5] Galinier, M; Pathak, A; Fourcade, J; Androdias, C; Curnier, D; Varnous, S; Boveda, S; Masabuau, P; Fauvel, M; Senard, JM; Bounhhoure, JP. Depressed low frequency power of heart rate variability as an independent predictor of sudden death in chronic heart failure. *Eur Heart J*, 2000, 21, 475-482.

[6] La Rovere, MT; Pinna, GD; Maestri, R; Mortara, A; Capomolla, S; Febo, O; Ferrari, R; Franchini, M; Gnemmi, M; Opasich, C; Riccardi, PG; Traversi, E; Cobelli, F. Short-Term Heart Rate Variability Strongly Predicts Sudden Cardiac Death in Chronic Heart Failure Patients. *Circulation*, 2003,107,514-516.

[7] Matveev, M and Prokopova, R. Diagnostic value of the RR-variability indicators for mild hypertension. *Physiol. Meas.*, 2002, 23, 671-682.

[8] Guzzeti, S; La Rovere, MT; Pinna, GD; Maestri, R Borroni, E; Porta, A; Mortara, A; Malliani, A. Different spectral components of 24 h heart rate variability are related to different modes of death in chronic heart failure. *Eur Heart J*, 2005, 26, 357-362.

[9] Pugsley, MK; Walker, MJA; Yong, SL. Are the arrhythmias due to myocardial ischaemia and infarction dependent upon the sympathetic system? *Cardiovascular Research,* 1999, 43, 830-831.

[10] Xiao-Jun, Du; Dart, AM. Role of sympathoadrenergic mechanisms in arrhythmogenesis. *Cardiovascular research*, 1999, 43, 832-834.

[11] Airaksinen, KE; Ylitalo, A; Niemela, M; Tahvanainen, KUO; Huikuri, HV. Heart rate variability and occurrence of ventricular arrhythmias during balloon occlusion of major coronary artery. *Am J Cardiol*, 1999, 83, 1000-1005.

[12] Shuterman, V; Aysin, B; Weiss, R; Brode, S; Gottipati, V; Schwartzman, D; Anderson, KP; for the ESVEM Trial Investigators. Dynamics of low-frequency R-R interval oscillations preceding spontaneous ventricular tachycardia. *Am Heart J*, 2000, 139, 126-133.

[13] Kong, TO; Goldberg, JJ; Parker, M; et al. Circadian variation in human ventricular refractoriness. *Circulation*, 1995, 92, 1507-1516.

[14] Adamopulos, S; Ponikovski, P; Cerquetani, E; Piepoli, M; Rosano, G; Sleight, P; Coats, AJ. Circadian pattern of heart rate variability in chronic heart failure patients. Effects of physical training. *Eur Heart J*, 1995, 1380-1386.

[15] Fries, R; Hein, S; Konig, J. Reversed circadian rhythms of heart rate variability and morning peak occurrence of sustained ventricular tachyarrhythmias in patients with implanted cardioverter defibrillator. *Med. Sci. Monit.*, 2002, 8, 751-756.

[16] La Rovere, MT; Mortara, A; Pantaleo, P; Maestri, R; Cobelli, F; Tavazzi, L. Scopolamine improves autonomic balance in advanced congestive heart failure. *Circulation*, 1994, 90, 838-843.

[17] Mortara, A; La Rovere, MT; Pinna, GD; et al. Nonselective beta-adrenergic blocking agent, Carvedilol, improves arterial baroreflex gain and heart rate variability in patients with stable chronic heart failure. *J Am Coll Cardiol*, 2000, 36, 1612-1618.

[18] Tjeerdsma, G; Szabo, BM; van Wijk, LM; Brouwer, J; Tio, RA; Crjins, HJGM; van Veldhuisen DJ. Autonomic dysfunction in patients with mild heart failure and coronary artery disease and the effects of add-on beta-blockade. *Eur J Heart Failure*, 2000, 3, 33-39.

[19] Packer, M; Bristow, MR; Cohn, JN; Colucci, WS; Fowler, MB; Gilbert, EM; Shusterman, NH; for the U. S. Carvedilol Heart Failure Study Group. The Effect of Carvedilol on Morbidity and Mortality in Patients with Chronic Heart Failure. *N. Engl. J. Med.*, 1996, 334, 1349-1355.

[20] Boudonas, G; Psirropoulos, DZ; Ginopoulos, D; Stampoulidis, K; Lefkos, N. Heart failure and effects of Carvedilol on heart rate variability, QT dispersion and ventricular arrhythmias. *Eur J Heart Failure Suppl*, 2004, 3, 119.

[21] Voronkov L, Bogachova N. Long-term Carvedilol restores autonomic drive of heart in stable chronic heart failure. *Eur J Heart Failure Suppl*, 2004, 3, 40.

[22] Prokopova, R; Matveev, M; Nachev, Ch. Correlating changes in heart autonomic balance and ventricular arrhythmias reflecting the positive effect of treating heart failure with Carvedilol. *Eur J Heart Failure Suppl*, 2005, 1, 37-38.

[23] Rabbia, F; Martini, G; Sibona, M; Grosso, T; Simondi, F; Chiandussi, L; Veglio, F. Assessment of Heart Rate Variability After Calcium Antagonist and Beta-Blocker Therapy in Patients wth Essential Hypertension. *Clin Drug Invest*, 1999, 17, 111-118.

SUMMARY OF THE RESULTS: CONCLUSIONS

Cardiovascular diseases are the most frequent cause of death in the world. The invalidization caused by them and the increasingly higher cost of treatment have transformed CVD into one of the most significant social problems. After clinical manifestation, the CVD patients are subjected predominantly to maintenance treatment. These negative facts determine the special importance of the studies clarifying the causes for the emergence and progression of CVD.

Knowledge of the principal factors leading to the appearance of cardiovascular pathology provides an opportunity for timely and appropriate help already before the clinical manifestation of CVD. The changes in the autonomic cardiac control attained through permanently changing balance between the two parts of the autonomic nervous system are among the first alarming signals for the onset of pathological processes in the cardiovascular system. The impaired autonomic balance is closely connected with the appearance and progression of cardiovascular diseases, and with the current and prognostic evaluation of cardiac risk.

Heart rate variability is a universally accepted method for assessing the autonomic control of the heart. At the same time, however, the prevalent opinion is that HRV is a highly *specific* method, but with a low *sensitivity* to the principal cardiovascular diseases. The results of our studies encourage us to assume that it is possible to obtain differentiated characteristics of the autonomic balance for concrete cardiovascular pathologies. We attained this by assessing the changes taking place in *the circadian characteristic* of the autonomic balance, which has a definite profile in norm and with the principal CVD. We also proposed an adequate method for evaluating the changes in the circadian nature of AB by comparing the values of the HRV indices during rest and upon VNS stimulation during two intervals in the 24-hour period in which there is a physiologically determined difference in the balance.

The normal circadian characteristic of AB is distinguished by a slight prevalence of the circadian nature of the parasympathetic component. This result is demonstrated using TRABI in healthy individuals. The mean value of the assessments of the HRV indices using TRABI, according to data from the study during rest and with the handgrip test is 0.100, according to data from the study during rest and with the Valsalva manoeuvre is 0.141. In almost all nosological groups studied by us – with the only exception of the group with unstable angina - the circadian nature has the opposite profile to the normal one. There is a prevalence of the circadian nature of the sympathetic component in the autonomic balance. The mean values of the scores of the indices for HRV using TRABI in the comparative studies during rest and with the two stimulation tests (handgrip test and Valsalva manoeuvre) are accordingly: with mild hypertension – 0.132 and 0.047; with morning MI incidents – 0.177 and 0.107; with non-morning MI incidents – 0.082 and 0.053; with HF – 0.276 and 0.219. The abnormal circadian characteristic of AB is evidence of the onset of cardiovascular pathology.

The results of the studies demonstrated that abnormal circadian nature is observed already with the appearance of CVD. We have detected changes in the circadian characteristic of AB in the mildest forms of the hypertensive disease.

The changes in the circadian nature of the autonomic balance are related also with the prognosis of CVD. The patients with unstable angina in the population studied have a good long-term prognosis. The circadian nature in them is with prevalence of the parasympathetic part in the autonomic balance. The mean values of the assessments of the HRV indices using TRABI in the case of unstable angina in the two comparative studies are 0.180 and 0.211. The fact that a near-normal circadian profile is associated with a better prognosis is also proven in patients with HF. After treatment of the heart failure with Carvedilol, we observed a tendency towards restoration of the autonomic balance. The mean values of the scores for the HRV indices using TRABI after treatment are 0.100 and 0.176 accordingly, i.e., the treatment has resulted in a change in the circadian profile in the direction of prevalence of the circadian nature of the vagal activity. The treatment with Carvedilol also reduces the risk of ventricular rhythm disorders, which are a principal cause of HRV in patients with CHF. These two results can be summarized with the conclusion that the restored circadian nature of AB during Carvedilol therapy in the case of HF indicates a better prognosis.

The hypothesis that the normal circadian nature of AB also has a preventive function against the development of damage to the target organs is also supported by the results of applying appropriate treatment for hypertension, modeling AB. By administering a blocker of the imidazoline receptors (Rilmenidine) to patients with mild hypertension, we have attained a practically complete restoration of the normal circadian profile. The mean values of the scores of the HRV indices using TRABI in mild hypertensive subjects after treatment are 0.90 and 0.139 accordingly.

In spite of the abnormal circadian nature of AB, which is common to CVD, we can claim that the circadian nature has a specific profile for every nosological unit. *The hypertension* is characterized by a slightly elevated circadian nature of the sympathetic component in AB and no circadian nature of the parasympathetic one. *Consequently, the principal disorder is relevant predominantly to the vagal activity.* Morning MI are characterized by strongly enhanced circadian nature of the sympathicus and almost normal parasympathetic circadian nature, i.e., *the principal disorder is in the sympathetic part of AB.* Non-morning MI are characterized by reduced circadian nature of the sympathicus and almost total absence of circadian nature of the parasympathicus. Obviously, *the principal disorder is relevant to the parasympathetic component in AB. The heart failure* is characterized by a strongly manifested circadian nature both of the sympathicus and of the parasympathicus. The mean values of the TRABI-scores are also the highest among the values for the CVD studied by us, and the circadian nature is abnormal. *Both AB parts are affected.* The disorder in the circadian profile in the case of *unstable angina* is also characteristic. The two AB parts have a stronger circadian nature, but the mean values of the scores of the HRV indices using TRABI do not differ significantly from those in healthy individuals. *The normal circadian profile is preserved.*

The results of the studies carried out suggest the following principal conclusions:

- the normal circadian nature of the autonomic balance is the principal characteristic of a healthy heart;
- cardiovascular diseases have a characteristic abnormal circadian nature of the autonomic balance;
- the preserved circadian nature with prevalence of the parasympathetic component is a reason for favorable prognosis in some of the most widespread CVD;
- the restoration of the normal circadian profile of AB through appropriate treatment constitutes also prevention against the onset of organ damage after CVD;
- the monitoring of the circadian nature is an adequate form of control on the results of the therapy.

The proposed Time-Related Autonomic Balance changes Indicator broadens, in our opinion, the opportunities for using HRV for HAB assessment. The method for evaluating the circadian changes in HAB through the HRV indices during rest and with stimulation during different time intervals is convenient and accessible for clinical use.

The characteristic circadian changes in AB in CVD should be taken into consideration when determining the therapeutic tactics. The therapy can be further specified in terms of the mechanism of action of the drug, which needs to be chosen in accordance with its capacity of correcting the principal AB disorder in a concrete patient. The distribution of the drug therapy over the 24-hour period should be specified in accordance with the circadian characteristic of AB. As the changes in AB and its circadian profile are among the causes for the onset of hypertensive disease, cardiac and cerebral vascular diseases and incidents, the therapeutic correction of these pathological changes can be included in the primary prevention of CVD.

INDEX

A

acetylcholine, 20, 21, 22, 23, 24, 108, 109
acid, 9, 16
action potential, 21, 24
activation, 3, 5, 10, 11, 23, 24, 27, 28, 43, 51, 78, 86, 127, 128, 129
adaptation, 5, 25, 127
adenosine, ix, 23, 49, 109
adhesion, 108
ADP, ix, 49
adrenaline, 17, 26, 30, 49, 54
adults, 50, 52
affect, 4, 20, 21, 23, 128
afternoon, 43, 53, 59, 60, 61, 62, 63, 64, 66, 67, 68, 69, 70, 71, 72, 73, 74, 81, 82, 83, 84, 85, 87, 88, 89, 90, 91, 93, 96, 97, 98, 99, 106, 111, 112, 113, 114, 116, 117, 118, 119, 120, 121, 129, 131, 132, 133, 134, 135, 136, 137, 138, 139, 140, 142
age, 49, 50, 57, 60, 77, 92
ageing, 50, 56
agent, 26, 144
aggregation, 49, 53, 54, 108, 109, 124
aging, 55
agonist, 101
aldosterone, xi, 26, 49, 86, 101, 105, 128
algorithm, 92, 93
alternative, 60
alternative hypothesis, 60
amplitude, 11, 33, 35, 37, 42, 50
amygdala, 17
angina, 17, 52, 54, 55, 57, 75, 104, 106, 107, 108, 109, 111, 116, 118, 122, 123, 124, 146
angiography, 54, 122
angioplasty, xi
angiotensin converting enzyme, 86
angiotensin II, ix, 45, 87, 101
animals, 4, 8, 10, 47, 48

ANS, ix, 4, 6, 23, 24, 27, 29, 40, 42, 50, 59, 112
antagonism, 19, 23, 24
antigen, 49, 109
antihypertensive drugs, 8, 86, 87
aorta, 5, 6, 23
argument, 40, 88
arousal, 123
arrhythmia, 39, 44, 103, 131
arterial hypertension, vii, 8, 9, 12, 60, 77, 78, 80, 98
arteries, 22, 26, 28, 40
arterioles, 42, 128
artery, ix, 29, 56, 74, 101, 104, 105, 108, 123, 129, 143
assessment, 59, 73, 74, 101, 147
association, 122, 123
atherosclerosis, 27, 78, 86, 106, 108
atria, 1, 2, 3, 4, 7, 19, 20, 21, 22, 24
atrial fibrillation, 21
atrioventricular node, 7, 30, 40
atrium, 21
attacks, 16, 124
attention, 43, 73
autonomic nervous system, ix, 4, 6, 15, 28, 29, 49, 50, 51, 57, 79, 81, 107, 110, 123, 145
autonomic neuropathy, 110, 121

B

baroreceptor, 8, 9, 10, 23, 42, 45
beta-blocker, 107, 110, 111, 130, 138
binding, 49
biological rhythms, 48
blood, ix, 2, 4, 5, 8, 10, 11, 12, 13, 17, 25, 26, 27, 28, 30, 40, 45, 54, 55, 56, 60, 78, 79, 92, 100, 106, 109, 123, 124
blood flow, 27, 106, 109
blood pressure, 8, 10, 13, 17, 28, 30, 45, 60, 78, 92, 100, 107, 109, 123

blood vessels, 8, 25, 26
body, 4, 48, 78
body mass index, 78
bradycardia, 10, 11, 13
bradykinin, 6
brain, 2, 3, 9, 11, 12, 14, 16, 17, 42, 48, 55, 57, 77, 79, 87
brain stem, 9, 14, 17, 42
breathing, 45
Bulgaria, vii, 1, 19, 33, 47, 59, 75, 77, 103, 127

C

calcium, 86, 87, 92, 111
cardiac activity, 8, 11, 12, 13, 14, 15, 20, 24, 40, 79
cardiac muscle, ix, 3, 7, 110
cardiac output, 51
cardiac risk, 59, 130
cardiomyopathy, 130
cardiovascular disease, ix, 27, 56, 59, 74, 77, 80, 110, 145, 147
cardiovascular risk, 50
cardiovascular system, vii, 6, 27, 40, 44, 48, 49, 56, 79, 129, 145
carotid sinus, 23
catecholamines, 8, 17, 25, 26, 27, 29, 30, 49, 50, 51, 53, 86, 105, 127, 128, 129
cell, 6, 8, 9, 124
central nervous system, ix, 6, 9, 12, 14, 16, 17
cerebral cortex, 8, 11, 12, 13, 14
changing environment, 48
chiasma, 48
children, 49
chromosome, 47
chronic renal failure, ix, 81
circadian rhythm, 43, 47, 48, 51, 53, 55, 56, 57, 74, 80, 100, 101, 107, 144
circadian rhythms, 47, 48, 55, 56, 57, 144
circulation, 5, 26, 28, 40, 55, 127
classification, 92, 93, 94, 95, 96, 97, 98, 99, 111
CNS, ix, 6, 7, 24
coagulation, 49, 79, 108
coenzyme, 105
collateral, 26, 109, 124
compliance, 92
complications, 27, 54, 99, 104, 111
components, 24, 42, 49, 50, 51, 54, 74, 109
composition, 74
computation, 70
concentration, 21, 22, 26, 86
concrete, 14, 61, 92, 145, 147
conduct, 3, 12, 81

conduction, 2, 4, 7, 19, 20, 22, 23, 24, 27, 28, 40, 81, 111, 115
conductivity, 20, 21, 23, 24
confidence, 60, 61, 94, 96
confidence interval, 94, 96
congestive heart failure, 8, 30, 104, 143, 144
control, vii, 5, 7, 8, 9, 11, 12, 14, 15, 17, 19, 20, 21, 27, 30, 40, 41, 42, 43, 44, 45, 54, 56, 59, 74, 78, 85, 98, 99, 101, 110, 115, 119, 135, 145, 147
cooling, 4, 16
coronary angioplasty, 105
coronary arteries, 7
coronary artery disease, 30, 122, 144
coronary heart disease, 104
correlation, 2, 3, 17, 25, 43, 44, 74, 105
correlation coefficient, 43
cortex, 9, 10, 11, 12, 13, 14, 15, 17
cortisol, 55, 105, 109
cranial nerve, 6, 8
creatine, ix, 110
CVD, ix, 27, 53, 112, 145, 146, 147
cycles, 34, 41

D

damage, 9, 13, 27, 29, 77, 79, 81, 100, 104, 106, 110, 111, 146, 147
data distribution, 99
death, xii, 13, 14, 15, 17, 53, 55, 56, 57, 103, 104, 106, 124, 128, 130, 143, 145
deduction, 99
defibrillator, 57, 144
deficit, 105
delivery, 78
demand, 54, 56, 106, 123
density, x, 21, 25
depolarization, 22, 41
deposits, 20, 21
depression, 12, 13, 17, 52, 98
deprivation, 48
detection, 103
deviation, x, xi, xii, 27, 34, 35, 69
diabetes, 30, 50, 110
diastole, 3
diastolic pressure, 2, 4
dilation, 4, 26, 108, 109
direct measure, 45
discharges, 2, 3, 4, 19, 21, 23
discriminant analysis, xi, 93
discrimination, 93, 94, 96, 97, 99, 102
disorder, 130, 146, 147
dispersion, xi, 21, 93, 130, 144

distribution, 1, 24, 34, 39, 52, 53, 60, 92, 110, 124, 147
diuretic, 4, 131
dogs, 21, 30, 45
domain, 34, 35, 36, 37, 38, 39, 43, 46, 60, 64, 66, 67, 70, 71, 72, 74, 78, 80, 81, 87, 98, 99, 103, 105, 122, 123, 128, 134, 135, 140
Drosophila, 47, 48, 55
drug therapy, 110, 111, 130, 147
drugs, 28, 87, 92, 100, 106, 111, 115, 130
duration, 7, 21, 23, 24, 33, 34, 39, 41, 44, 53, 64, 99, 105, 110

E

efferent nerve, 27, 42, 45
elasticity, 73, 83
electrocardiogram, x, 12, 30, 33, 122
emergence, 8, 106, 145
emotions, 10, 14
endocardium, 22, 23
endocrine, 41
endothelium, 108, 109
environment, 47, 48
enzymes, 21
epicardium, 2, 6, 22, 23
epilepsy, 10, 13
epinephrine, 109
equality, 61
eukaryote, 47
evening, 48, 52, 53, 110
evidence, 13, 42, 47, 53, 77, 81, 87, 98, 99, 103, 104, 110, 111, 129, 133, 146
evolution, 48
examinations, 99, 115
excitability, 2, 10, 26
excitation, 2, 4, 14, 15, 20, 26, 33, 51
exclusion, 111, 115
excretion, 128
exercise, 4, 21, 23, 57, 81, 103, 122

F

failure, 16, 17, 50, 92, 128, 129, 130, 131, 144
false negative, 95, 97, 98
false positive, 95, 97, 98
Fast Fourier transform, 36
fatal arrhythmia, 53, 129
feedback, 7, 10, 40
feelings, 11
FFT, x, 36, 38, 39
fibers, 1, 2, 3, 4, 5, 7, 8, 12, 16, 20, 21, 22, 25, 42

fibrillation, 7, 11, 12, 24
fibrinogen, 49
fibrinolysis, 49, 53, 54, 79, 108, 109
fluctuations, 5, 40, 41, 42, 48, 50, 80
fluid, 4, 26
food, 9
foramen, 6
free radicals, 86, 105, 124
freezing, 5
frontal cortex, 13, 14
frontal lobe, 14
functional changes, 40

G

ganglion, 3, 6, 129
gender, 55, 92
gene, 33, 55
gene transfer, 55
generation, 11, 27
genes, 47, 48, 55
gland, 8
glucose, 78, 105
glutamate, 8, 9, 11
gray matter, 8, 11, 12, 14
groups, 8, 9, 29, 50, 52, 54, 73, 79, 80, 92, 93, 94, 95, 96, 97, 99, 104, 107, 109, 111, 112, 115, 129, 145

H

heart attack, 16
heart disease, 56, 60, 123, 124
heart failure, vii, ix, x, 27, 28, 57, 104, 110, 111, 115, 127, 128, 129, 130, 133, 143, 144, 146
heart rate, vii, x, 5, 7, 8, 10, 16, 17, 20, 21, 23, 24, 28, 29, 30, 33, 34, 39, 40, 41, 42, 44, 45, 48, 49, 51, 52, 55, 56, 57, 74, 75, 77, 79, 98, 100, 101, 105, 106, 110, 122, 123, 125, 128, 143, 144
hippocampus, 12
homeostasis, 8
homogeneity, 130
hormone, 79
hospitalization, 103
hyperactivity, 27, 28, 77, 78, 102
hypersensitivity, 24, 25
hypertension, xi, 9, 14, 16, 17, 27, 28, 30, 46, 50, 51, 54, 56, 74, 77, 78, 79, 80, 81, 85, 86, 87, 92, 94, 95, 97, 98, 99, 100, 101, 102, 103, 111, 143, 146
hypertrophy, xi, 8, 27, 28, 30, 43, 46, 51, 56, 78, 81, 86, 100, 101
hypotension, 13

hypotensive, 11
hypothalamus, 8, 9, 10, 11, 12, 13, 14, 15, 43
hypothesis, vii, 14, 70, 81, 92, 94, 96, 98, 106, 115, 121, 135, 146

I

identification, 78, 128
identity, 94, 96
immobilization, 128
incidence, 52, 53, 56, 77, 79, 107, 129
inclusion, 98, 99
independence, 122
indicators, 29, 49, 93, 94, 96, 98, 99, 102, 143
indices, vii, 33, 34, 35, 40, 43, 44, 59, 60, 61, 62, 63, 64, 65, 66, 67, 68, 70, 71, 72, 73, 74, 78, 79, 80, 81, 82, 83, 84, 85, 86, 87, 88, 89, 90, 91, 92, 93, 94, 101, 103, 104, 105, 107, 109, 111, 112, 113, 114, 115, 116, 117, 118, 119, 120, 121, 124, 128, 129, 130, 131, 132, 133, 134, 135, 136, 137, 138, 139, 140, 141, 142, 145, 146, 147
indirect effect, 27
infarction, 25, 52, 54, 103, 104, 115, 121, 129, 143
inflammation, 109
influence, 4, 5, 7, 8, 12, 14, 20, 21, 24, 25, 27, 39, 41, 43, 57, 60, 68, 74, 78, 81, 86, 87, 105, 106, 110, 125, 130, 137, 141
inhibition, 5, 9, 10, 20, 28, 43, 54, 86
inhibitor, xi, 49, 109, 131
inhomogeneity, 21
insulin, 55
integration, 9, 14
integrity, 13
intensity, 11, 20, 23, 42, 132
interaction, 7, 8, 12, 14, 23, 24, 45
interactions, 3, 5, 9, 29
interest, 25, 33, 39, 77, 86
interneurons, 3
interpretation, 44, 74
interrelations, 26
interval, 7, 11, 12, 13, 20, 22, 30, 33, 34, 35, 38, 41, 43, 44, 60, 73, 110, 130, 143
intervention, 77, 130
intima, 28, 51, 52
inversion, 50, 130
ischaemic heart disease, vii, x, 27, 53, 103, 104
ischemia, 56, 123
isolation, 48

K

kidneys, 77, 79

knowledge, 12

L

latency, 23
lead, 11, 15, 22, 39, 53, 54, 129, 140
left atrium, 21
left ventricle, 2, 5, 22
lesions, 9, 16, 17, 106
limbic system, 13, 14
limitation, 37, 68
links, 3, 9, 10, 11, 13, 14
lipids, 78
lipoproteins, x
liquids, 4
liver, 78
localization, 4, 105
location, 1, 22
locus, 9
long run, 42
lumen, 25
lying, 10, 42, 43, 49, 108

M

management, 45
mass, 78, 101
measurement, 44, 60, 68, 78, 79, 93, 139
measures, 46, 68, 122, 123
median, 6, 11, 34, 48
medication, 92
medulla, ix, xii, 1, 4, 6, 8, 9, 10, 12, 13, 16, 17
melatonin, 79
men, 50, 52, 60, 78, 92, 124
mental activity, 53, 106
mercury, 28
mesencephalon, 11, 17
metabolic disorder, 127
metabolic syndrome, 86
metabolism, 27
metabolites, 6
mild hypertensive, 77, 81, 83, 85, 87, 88, 89, 90, 92, 98, 99, 146
modeling, 146
models, 14, 42, 97, 99
monitoring, 48, 50, 52, 79, 81, 108, 123, 147
morning, xi, 43, 48, 49, 52, 53, 54, 55, 56, 57, 59, 60, 61, 62, 63, 64, 65, 66, 67, 68, 69, 70, 71, 72, 73, 74, 75, 80, 81, 82, 83, 84, 85, 87, 88, 89, 90, 91, 93, 95, 97, 98, 99, 106, 107, 108, 109, 110, 111, 112, 113, 114, 115, 116, 117, 118, 119, 120,

121, 123, 124, 129, 131, 132, 133, 134, 135, 136, 137, 138, 139, 140, 141, 142, 144, 146
morphology, 54
mortality, 103, 104, 122, 127, 128, 130
motivation, 11, 81
multidimensional, 99
multiple regression, 106
muscles, 4, 25, 26, 29, 40, 78
myocardial infarction, xi, 8, 17, 23, 25, 27, 52, 53, 56, 57, 92, 103, 104, 109, 110, 111, 115, 122, 123, 124, 125, 129
myocardial ischemia, xii, 16, 57, 123
myocardium, 1, 2, 3, 5, 7, 13, 21, 22, 23, 26, 53, 54, 106, 109, 110, 124, 129, 130

N

necrosis, 128
needs, 147
nerve, vii, 1, 3, 4, 5, 7, 10, 11, 13, 14, 16, 19, 20, 21, 22, 23, 25, 40, 41, 45, 50, 51, 56, 100, 101, 129
nerve fibers, 1, 5, 7, 11, 22, 23, 40
nervous system, xii, 13, 14, 33, 40, 49, 74, 77, 86, 98, 99, 100, 101, 130
network, 1, 7
neurological disease, 60, 92
neurons, 1, 6, 7, 8, 9, 10, 12, 14, 15, 17, 42, 44, 48
neuropathy, 49, 50, 110
neuropeptides, 23
neurotransmitter(s), 8, 10, 11
nitrates, 107
nitrogen, 105
N-N, xi, 34
non-smokers, 52
norepinephrine, 30, 56, 143
normal distribution, 60
North America, 33, 44
nucleus(i), xi, 1, 3, 6, 8, 9, 10, 11, 12, 13, 16, 17, 48
null hypothesis, 60, 61

O

obesity, 101
observations, 60
obstruction, 4
occlusion, 29, 129, 143
organ, 27, 42, 60, 77, 147
organism, 26, 40, 47, 48, 77, 79, 127
organization, 14, 17, 47, 50
oscillation, 35, 36, 45
ostium, 20, 21
outline, 10

output, 21, 60
oxidative stress, 28, 101
oxygen, 54, 106, 108, 110
oxygen consumption, 110

P

pain, 52
parameter, 61, 74
passive, 47
pathogenesis, 13, 14, 22
pathology, vii, 33, 48, 145, 146
pathophysiology, 79, 81, 109, 129
pathways, 1, 4, 9, 12, 19, 22, 27, 28, 29, 40
perfusion, 27, 54
periodicity, 41, 42, 47, 107
peripheral vascular disease, 110
phosphorylation, 23
physical activity, 52, 56, 107, 110, 123, 128
physics, 75
physiological correlates, 40
placebo, 53, 106, 107, 110
plasma, 4, 13, 17, 26, 29, 49, 51, 54, 55, 79, 86, 109, 128, 130
plasma levels, 4, 13, 51, 54
plasminogen, xi, 49, 109
plasticity, 16
platelet aggregation, 49, 53
platelets, 53, 54, 108, 109
plexus, 1, 6, 7
PM, 17, 56, 124
polarity, 12
pons, 17
poor, 75, 103, 104, 122, 127
population, 9, 92, 104, 112, 114, 115, 146
posterior cortex, 13
posture, 5, 55, 123
power, 36, 42, 43, 44, 51, 56, 60, 68, 72, 73, 74, 93, 97, 98, 99, 101, 104, 130, 143
prediction, 40, 78
predictors, 103, 104
prefrontal cortex, 13
preparation, 39
pressure, ix, x, xii, 2, 3, 4, 5, 8, 11, 12, 13, 14, 17, 23, 26, 27, 28, 29, 42, 45, 48, 49, 50, 52, 54, 77, 78, 79, 80, 81, 86, 92, 100, 101, 106, 110
prevention, 106, 130, 147
principle, 43
probability, 51, 52, 53, 60, 105
probe, 45
production, 8, 26
prognosis, 27, 78, 100, 103, 104, 106, 111, 112, 115, 122, 146, 147

program, 101
propranolol, 11
protective factors, 105
protective role, 115, 122, 129
psychological stress, 53, 129
pulse, 28, 29, 48

R

range, 35, 41, 42, 43, 44, 59, 69, 79, 92, 93, 98, 115
reagents, 109
recall, 121, 137
receptors, 1, 2, 4, 5, 6, 7, 8, 11, 14, 16, 22, 26, 86, 87, 127, 146
recovery, 16, 105
reduction, 2, 5, 9, 23, 24, 39, 43, 50, 51, 53, 54, 59, 70, 79, 80, 81, 86, 87, 88, 106, 107, 109, 112, 128, 129, 138, 140
reflection, 49, 83, 87, 130, 138, 140
reflexes, 3, 4, 10, 14
regression, 100
regulation, 4, 5, 6, 7, 8, 9, 10, 11, 12, 13, 14, 16, 23, 24, 25, 27, 43, 44, 78, 86, 108, 127, 130
relationship, 73
reliability, 43
remodelling, 14, 28
renal failure, 111
renin, xi, 5, 26, 27, 28, 43, 49, 54, 55, 86, 98, 101, 105, 128
resection, 13
resistance, 4, 5, 10, 26, 28, 45, 77, 79, 109, 127
resolution, 44
respiration, 39, 41, 48, 127, 128
respiratory, xi, 8, 17, 28, 39, 41, 44, 45
rhythm, 3, 8, 10, 11, 13, 19, 20, 22, 41, 42, 47, 48, 55, 57, 104, 107, 125, 128, 140, 146
rhythmicity, 47
risk, 20, 53, 54, 57, 59, 70, 74, 77, 78, 79, 80, 86, 87, 99, 103, 104, 106, 109, 110, 111, 124, 128, 129, 130, 141, 145, 146
risk factors, 74, 77, 103, 109
room temperature, 43

S

sample, 36, 112, 114, 115, 120
SAP, xii, 92
saturation, 115
scores, 145, 146
searching, 60
secretion, 4, 5, 7, 9, 17, 43, 49
self, 47

sensitivity, vii, 1, 3, 26, 70, 71, 72, 74, 75, 81, 104, 138, 145
septum, 13, 22
series, 13
serotonin, 8, 105
severity, 59, 77, 81, 104, 108, 127, 128, 131
shear, 110
sign, 60, 61, 64, 66, 69, 75, 104
signals, 4, 36, 37, 44, 48, 145
similarity, 48
sinus, xi, 4, 7, 11, 19, 20, 21, 22, 23, 24, 30, 33, 39, 40, 41, 42, 45, 81, 111, 128, 130
sinus arrhythmia, xi, 39, 45
sinus rhythm, 81, 111, 130
sites, 1, 4, 14
smokers, 52
social problems, 145
sodium, 26, 28, 56, 77
specificity, vii, 2, 44, 69
spectral component, 42, 45, 54, 75, 128, 143
spectrum, 36, 40, 41, 43, 45, 51, 75
speed, 5
spinal cord, 3
spine, 2, 3, 6, 8, 11, 12, 14
stability, 70, 73, 74, 114, 137
stabilization, 99, 135
stable angina, 57, 106, 107, 109, 123
stages, 78, 81
standard deviation, 34, 35
standardization, 33, 39, 93
standards, 43
stellate, 129
stimulus, 2, 6
strategies, 105
stratification, 30, 103, 106, 111, 122
stress, 13, 14, 26, 29, 70, 79, 88, 89, 109, 110, 112, 115, 116, 118, 124, 132, 141
stretching, 3
stroke, 7, 16, 55, 77, 79, 110, 127
stroke volume, 7, 77, 127
substitution, 93
supply, 54, 106, 108
suppression, 28, 86, 87, 90, 102, 136, 137, 139, 141
suprachiasmatic nucleus, 48
surface layer, 22
susceptibility, 14, 49, 50, 52
symmetry, 69, 70
sympathetic denervation, 20, 22, 25, 29
sympathetic fibers, 3, 7, 42
sympathetic nerve fibers, 7, 22
sympathetic nervous system, 2, 11, 27, 52, 77, 127
symptoms, 11, 52, 56, 110
synaptic transmission, 11

synchronization, 48
syndrome, 22, 29, 51
synthesis, 93, 94, 96, 127
systems, vii, 48, 50, 79, 128

T

tachycardia, 4, 9, 10, 11, 12, 13, 19, 127, 140, 143
tactics, 86, 147
target organs, 27, 77, 79, 81, 87, 146
tau, 47
teeth, 33
temperature, 48
temporal lobe, 12, 13
tension, 25
terminals, 23
thalamus, 11, 13
theory, 93
therapy, 29, 77, 81, 86, 87, 88, 89, 90, 92, 103, 105, 106, 107, 109, 110, 111, 115, 130, 131, 135, 136, 137, 138, 139, 140, 141, 146, 147
threshold, 7, 24, 26, 41
thrombus, 106, 108
time, 3, 11, 13, 16, 19, 22, 23, 24, 27, 28, 33, 34, 35, 36, 37, 38, 39, 41, 42, 43, 44, 46, 47, 53, 54, 57, 59, 64, 66, 67, 68, 69, 70, 71, 72, 73, 74, 78, 79, 80, 81, 86, 87, 89, 98, 99, 103, 104, 107, 108, 110, 113, 114, 115, 116, 120, 122, 128, 130, 132, 133, 135, 137, 138, 140, 145, 147
time series, 34
tissue, xii, 7, 21, 23, 25, 27, 48, 49, 109, 127
tissue plasminogen activator, xii, 49, 109
tonic, 4, 8, 9, 23
traffic, 3, 4, 39, 129
training, 57, 93, 97, 99, 143
transformation, 35, 36, 37, 38
transition, 42, 73
transmission, 16
transplantation, 48
trend, 98, 99
trial, 56
triggers, 56
tumor, 48

U

unstable angina, xii, 105, 108, 111, 114, 115, 118, 121, 123, 145, 146
urine, 86

V

vagus, 1, 4, 6, 10, 19, 22, 23, 24, 28, 41, 43, 70, 112, 117, 118, 121, 132, 134
values, 34, 35, 36, 38, 43, 44, 49, 50, 51, 54, 59, 60, 61, 64, 66, 67, 68, 69, 70, 73, 78, 79, 80, 81, 82, 83, 84, 85, 86, 87, 88, 89, 90, 92, 93, 94, 95, 97, 98, 99, 104, 105, 106, 107, 111, 112, 113, 114, 116, 117, 118, 119, 120, 121, 128, 129, 130, 131, 132, 133, 134, 135, 136, 137, 138, 139, 140, 141, 145, 146
variability, vii, 16, 29, 30, 33, 34, 39, 40, 45, 46, 49, 50, 55, 56, 57, 73, 74, 75, 78, 90, 100, 101, 102, 122, 123, 125, 140, 143, 144, 145
variables, 30, 99, 122
variant angina, 108, 109
variation, 39, 44, 45, 49, 55, 56, 57, 74, 99, 108, 123, 124, 143
vascular diseases, 147
vasoconstriction, 26, 29, 52, 54, 108, 109
vasodilation, 26, 108
vasomotor, 2, 3, 4, 5, 9, 17, 42, 44, 108, 124
vasopressin, 4, 5, 9
vasospasm, 108
velocity, 2, 20
ventricle, 2, 3, 4, 22
ventricular arrhythmias, 25, 30, 103, 124, 128, 140, 141, 143, 144
ventricular tachycardia, 11
vessels, 1, 4, 5, 25, 26, 27, 42, 51, 54, 78, 79, 105, 106, 108, 109, 110, 127
viscosity, 49, 54, 55, 79, 109, 124
vulnerability, 104, 129, 130

W

waking, 17, 49, 107
water, 28
wear, 36
white blood cells, 109
women, 50, 52, 60, 92
work, 33
workers, 28

Y

young men, 50

SOUTH UNIVERSITY LIBRARY

BAKER & TAYLOR

Matveev, Mikhail.
Normal and abnormal
circadian
characteristics in
autonomic cardiac

QP
113
.N667
2006

48703
$89.00